Southern Living

Basic Plumbing

Oxmoor House®

Southern Living® Basic Plumbing was adapted from a book by the same title published by Sunset Books.

Consulting Editors: Don Vandervort, Richard Day
Editorial coordinator: Vicki Weathers

Staff for this book:
Senior Editor: Heather Mills
Senior Art Director: Francine Lemieux
Art Director: Normand Boudreault
Editor: Alfred LeMaitre
Assistant Editors: Caroline Bowden, Rebecca Smollett
Designers: François Daxhelet, Hélène Dion,
 Jean-Guy Doiron, François Longpré
Research Assistant: Adam van Sertima
Picture Editor: Christopher Jackson
Contributing Illustrators: Michel Blais, Jacques Perrault
Production Manager: Michelle Turbide
System Coordinator: Eric Beaulieu
Photographer: Robert Chartier
Proofreader: Elizabeth Warwick
Indexer: Christine M. Jacobs
Administrator: Natalie Watanabe
Other Staff: Lorraine Doré, Michel Giguère,
 Solange Laberge

Cover: Design by James Boone and Vasken Guiragossian. Photography by Norm Plate. Photo styling by Jean Warboy.

Our appreciation to the staff of *Southern Living* magazine for their contributions to this book.

Acknowledgments
Thanks to the following:
American Standard, Piscataway, NJ
American Water Heater Group, Johnson City, TN
Cast Iron Soil Pipe Institute, Chattanooga, TN
Centre Do-It D'Agostino, Montreal, Que.
Delta Faucet Co., Indianapolis, IN
Genova Inc., Davison, MI
Gerber Plumbing Fixtures Corp., Chicago, IL
Merle Henkenius, Lincoln, NE
John Kerns Heating and Plumbing, Millis, MA
Kohler Canada Inc., Etobicoke, Ont.
Moen Canada Inc., Oakville, Ont.
Sears-Roebuck, Chicago, IL
Watertown Public Works, Watertown, NY
Waxman Industries, Bedford Heights, OH
Whirlpool Corp., Benton Harbor, MI

First printing January 1999
Copyright © 1999 by Oxmoor House, Inc.
Book Division of Southern Progress Corporation
P.O. Box 2463, Birmingham, Alabama 35201
All rights reserved, including the right of reproduction in whole or in part in any form.

Southern Living® is a federally registered trademark of Southern Living, Inc.

ISBN 0-376-09053-7
Library of Congress Catalog Card Number: 98-87003
Printed in the United States

CONTENTS

4 GETTING STARTED IN PLUMBING
5 Your Plumbing System
8 Tools of the Trade
10 Working Safely
11 Working With Plastic Pipe and Tube
17 Working With Copper Tube
22 Working With Galvanized Steel Pipe
25 Working with Cast-Iron Pipe

27 PLUMBING REPAIRS
28 Clearing Out Clogs
33 Fixing Faucets
45 Solving Other Sink Problems
48 Toilet Repairs
57 Pipe Problems
61 Trap Problems
63 Stopping Valve Leaks
65 Water-Pressure Ups and Downs

66 PLUMBING IMPROVEMENTS
67 Roughing-In and Extending Pipe
75 Adding Shutoff Valves
77 Installing Faucets and Shower Heads
81 Putting in Sinks
85 Replacing Toilets
88 Installing Appliances

95 GLOSSARY

96 INDEX

GETTING STARTED IN PLUMBING

In this chapter you'll discover the inner workings of your plumbing system. Although everything is hidden behind walls or under floors, this network of pipes is not as complicated as you may think; the illustration opposite shows how the supply, drain-waste, and vent systems work together to carry water in and out of your home.

You'll also learn about the common tools and supplies required for the plumbing repairs and improvements shown in this book and learn the pipefitting basics for plastic, copper, galvanized steel, and cast iron. You'll need to know this for some of the plumbing improvements starting on page 66. If your projects are less ambitious, you'll find basic plumbing repairs in the chapter that starts on page 27.

For plumbing terms, refer to the glossary at the back. If you're dealing with plastic or copper, an important distinction to keep in mind is the difference between "pipe" and "tube." Pipe, with its thicker walls, is sized according to iron-pipe sizing, whereas thinner-walled tube is based on copper water-tube sizes.

Play it safe by reading the safety guidelines on page 10. Be sure to learn how to turn off the water supply—required in many plumbing jobs—also shown on page 10.

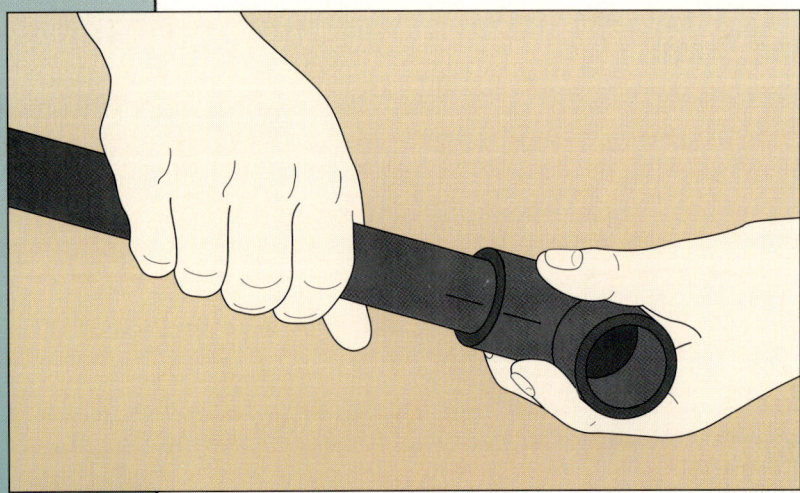

Most pipefitting jobs, such as fitting an elbow to change the angle of run, are easy once you have the know-how. Turn to page 11 to learn more.

YOUR PLUMBING SYSTEM

If your plumbing experience has up to now been limited to turning a faucet on and off, you'll be surprised at the simplicity of the system of pipes behind that faucet. Actually, as shown below, there are three separate but interdependent systems: supply, drain-waste, and vent. (Drain-waste and vent systems are interconnected and are referred to as the DWV system.) Before you begin any plumbing project, large or small, it's a good idea to become familiar with these systems. Once you understand how plumbing works, you'll find that

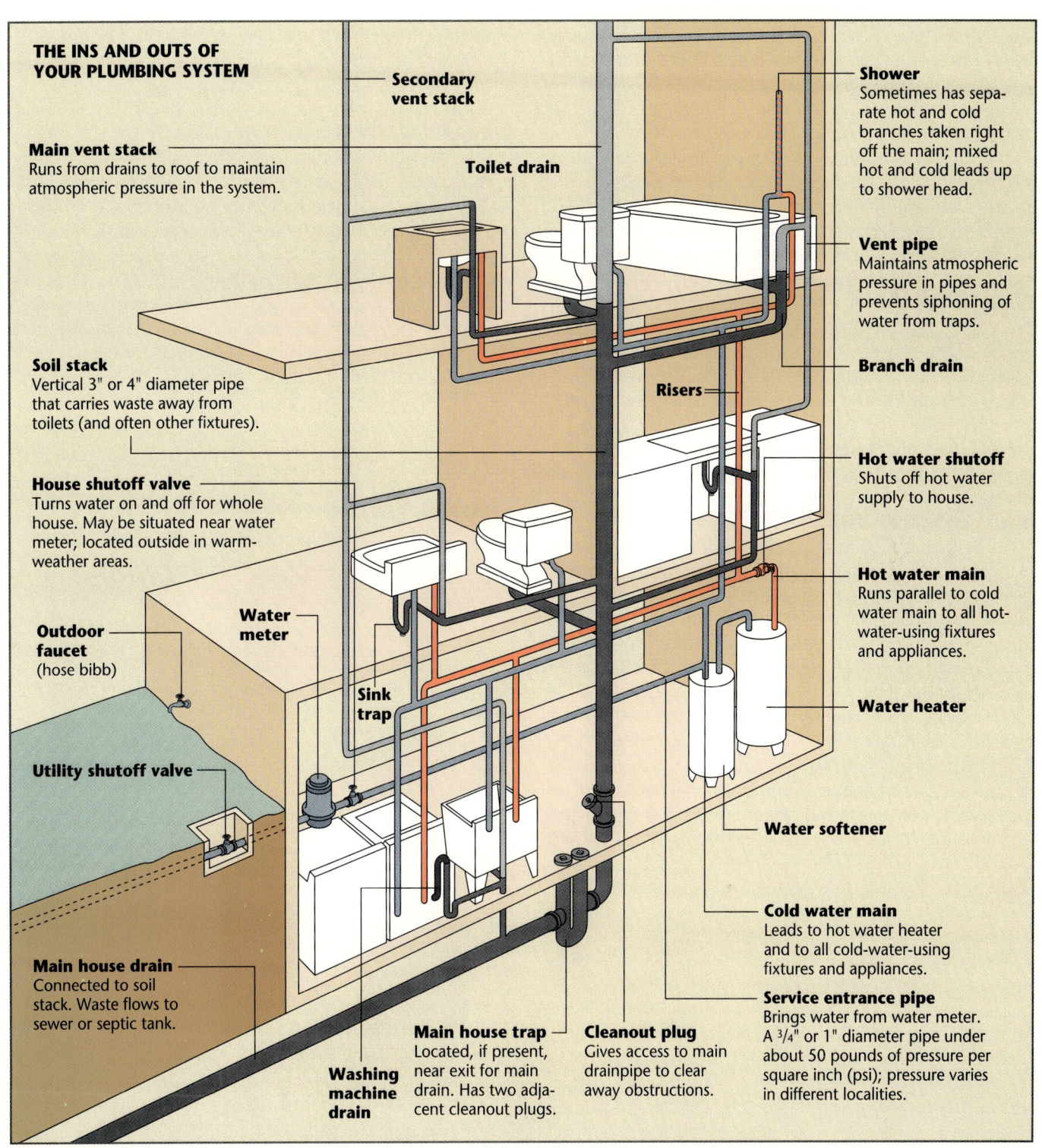

THE INS AND OUTS OF YOUR PLUMBING SYSTEM

Secondary vent stack

Main vent stack
Runs from drains to roof to maintain atmospheric pressure in the system.

Toilet drain

Shower
Sometimes has separate hot and cold branches taken right off the main; mixed hot and cold leads up to shower head.

Vent pipe
Maintains atmospheric pressure in pipes and prevents siphoning of water from traps.

Soil stack
Vertical 3" or 4" diameter pipe that carries waste away from toilets (and often other fixtures).

Risers

Branch drain

House shutoff valve
Turns water on and off for whole house. May be situated near water meter; located outside in warm-weather areas.

Hot water shutoff
Shuts off hot water supply to house.

Hot water main
Runs parallel to cold water main to all hot-water-using fixtures and appliances.

Outdoor faucet
(hose bibb)

Water meter

Sink trap

Water heater

Utility shutoff valve

Water softener

Main house drain
Connected to soil stack. Waste flows to sewer or septic tank.

Washing machine drain

Main house trap
Located, if present, near exit for main drain. Has two adjacent cleanout plugs.

Cleanout plug
Gives access to main drainpipe to clear away obstructions.

Cold water main
Leads to hot water heater and to all cold-water-using fixtures and appliances.

Service entrance pipe
Brings water from water meter. A 3/4" or 1" diameter pipe under about 50 pounds of pressure per square inch (psi); pressure varies in different localities.

GETTING STARTED IN PLUMBING

making repairs or adding fixtures is nothing more than a series of logical connections.

THE SUPPLY SYSTEM

If you get your water from a water utility, it's probably delivered by an underground water main through a utility shutoff valve and a water meter. If your water is provided by a water utility but is not metered, the utility shutoff valve is likely to be at your property line; if you can't locate it, check with the water utility. Where water comes from a private well, the shutoff valve is usually located where the water supply line enters the house, or at the wellhead—or both. This valve should be a full-flow type—usually a gate valve—so it won't restrict water flow into the house. To prevent freeze-ups, water lines are buried below the frostline.

Once inside the house, the supply pipe branches out into pipes of smaller diameters to deliver water to all fixtures and water-using appliances. (If the water in your house is softened or filtered, the treatment units will be attached near the point where the water enters the house.) A pipe (usually 3/4-inch), called a cold water main, leads to the hot water heater and to all cold-water-using fixtures and appliances. A hot water main begins at the water heater and runs parallel to the cold water main to all hot-water-using fixtures and appliances. Hot and cold branches, usually 1/2-inch pipe, serve fixtures such as the kitchen sink, or groups of fixtures, such as those found in the bathroom. A shower, because of its need for temperature control, will ideally get its own separate hot and cold water branches taken right off the main.

Hot and cold water mains are usually 3/4 inch in diameter. Branches that feed fixtures are generally 1/2-inch galvanized iron, copper, or plastic pipe. Local codes and the age of your house will affect the kinds of pipe and fittings you'll find and will determine what you can use if you're planning any changes or additions. (See the section on pipefitting beginning on page 11.)

Pipes that run vertically from one floor to the next are called risers. Long risers are often supported at their bases by platforms and anchored to wall studs. Horizontal runs are generally secured to floor joists or wall studs.

Supply pipes are installed with a slight pitch in the runs, sloping back toward the lowest point in the system so that all pipes can be drained. At the lowest point, there is sometimes a stop-and-waste valve that can be opened to drain the system.

Most fixtures and water-using appliances have their own shutoff valves to enable you to work at one place without cutting off the water supply for the entire house. (For instructions on how to install fixture shutoff valves, see page 75.) To be prepared for an emergency, everyone in the household should learn how to turn off the water supply, both at individual fixtures and at the house shutoff valve *(page 10).*

THE GAS AND HEATING SYSTEM PIPES

If you're planning to do a plumbing job yourself, you must be able to distinguish the water supply pipes from the pipes that carry gas into your home for household appliances or a water heater. A gas pipe is usually a black pipe that runs from the gas meter directly to a gas appliance or heating system. A separate shutoff valve for emergencies is required on each gas supply pipe. Don't try to work on gas piping yourself; call a professional.

Heating system pipes demand equal caution as well. To locate your heating pipes (either hot water or steam), trace them between each heating outlet and the furnace or other heat source. And by all means leave repairs to a heating expert.

THE DRAIN-WASTE SYSTEM

Unlike the supply system, which brings in water under pressure, the drain-waste system carries used water and waste out of the house (with the help of gravity) via the main house drain to the sewer or septic tank. These pipes lead away from all fixtures at a precise slope. If the slope is too steep, water will run off too fast, leaving solids behind; if it's not steep enough, water and waste will drain too slowly and may eventually back up into the fixture. The normal pitch is 1/4 inch for every horizontal foot of pipe. Cleanout fittings, as shown below, provide easy access to horizontal drainpipes, which clean out obstructions.

CLEANOUT FITTINGS

Y-fitting with cleanout insert

T-fitting

THE VENT SYSTEM

The vent system gets rid of sewer gas and prevents pressure buildup in the pipes by maintaining atmospheric pressure inside the system. To prevent dangerous sewer gases from entering the home, each fixture must have a trap which must be vented. A trap is a bend of pipe that is filled with water at all times to keep gases from coming up the drains. The two main types are P-traps *(far right)* and, in older homes, S-traps *(near right)*; the latter are no longer allowed by plumbing codes.

Usually, a house has a 3- or 4-inch-diameter main vent stack with 1½- to 2-inch branch vent pipes connecting to it. Branch vents are connected to the stack above the highest fixture that drains into the stack. When drainage-type fittings are used to make the stack connections, they are inverted to route the vent flow upward. In single-story homes, widely separated fixtures make a single main vent stack impractical, so each fixture or fixture group has its own waste connection and its own main or secondary vent stack.

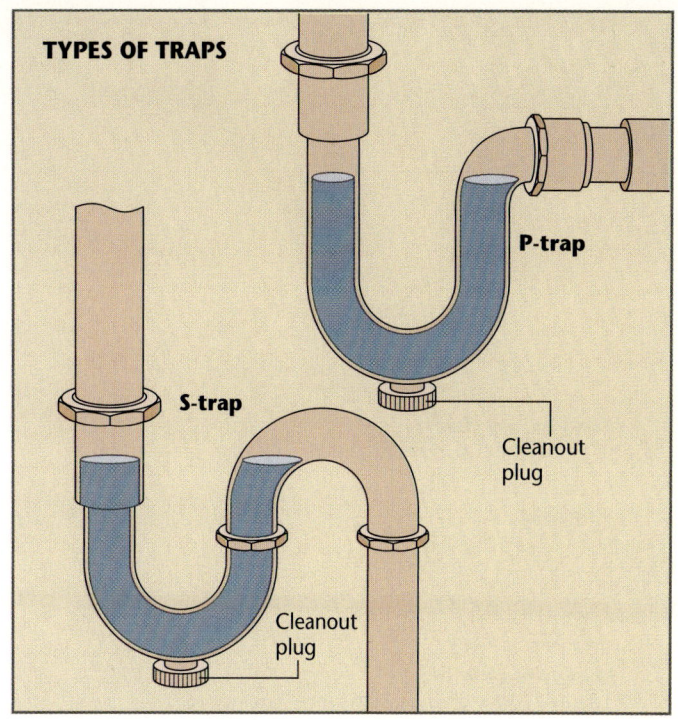

TYPES OF TRAPS

HOW TO READ YOUR WATER METER

By learning to read your water meter, you'll be able to easily keep track of water usage in your home, as well as to detect any leaks in the system.

Your home's water usage is probably measured by one of three meters. The six-dial meter *(below, left)* is by far the most common. Five of its six dials (labeled 10, 100, 1000, 10,000, and 100,000 for the number of cubic feet of water they record per revolution) are divided into tenths. The needles of the 10,000 and 100 meters move clockwise, and the other three move counterclockwise. The remaining dial (usually undivided), measures a single cubic foot per revolution. To read the six-dial meter, begin with the 100,000 dial, noting the smaller of the two numbers nearest the needle. Then read the dial labeled 10,000 and so on. This meter reads 628,260 cubic feet.

The five-dial meter *(below, middle)* is read in exactly the same way as the six-dial meter, except that single cubic feet are measured by a large needle that sweeps over the entire face of the meter. The meter in this example reads 458,540 cubic feet.

The digital-readout meter *(below, right)* looks like an automobile odometer. This type of meter may also have a small dial that measures a single cubic foot per revolution.

You can keep track of the water used by a specific appliance by simply subtracting the "before" reading from the "after" reading on your meter. To track down a possible leak, turn off all the water outlets in the house and note the position of the one-foot dial on your meter. After 30 minutes, check the dial. If the needle has moved, you have a leak.

Water meter
Found near the house shutoff valve; in the basement or crawl space in cold-winter areas, or else near the property line.

GETTING STARTED IN PLUMBING 7

TOOLS OF THE TRADE

PLUMBING TOOLS

Pipe cutter
Designed to cut through copper, galvanized steel, or cast-iron pipe.

Valve-seat wrench
Its hexagonal and square ends remove and replace worn or damaged valve seats.

Plastic tubing cutter
Cuts through flexible and rigid plastic tube.

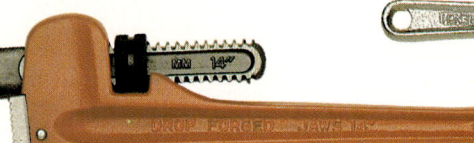

Spud wrench
Adjusts to fit nuts up to 4", such as slip nuts on traps and tailpieces.

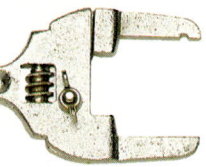

Valve-seat dresser
Grinds and smooths faulty non-replaceable valve seats in old faucets. Has various-sized cutters.

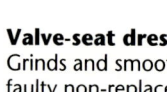

Pipe wrench
Toothed jaws grip pipe; some jobs require two wrenches. Available in sizes from 12" to 18" and larger; choose the proper size wrench for the pipe. Can use on nuts when an adjustable wrench fails, but will tend to damage the nut.

Basin wrench
Gives access to nuts behind sinks and other hard-to-reach places.

Adjustable wrench
Has smooth jaws made to fit small nuts, bolts, and square and hexagonal fittings. Use a 10" or 12" wrench.

Long-nose pliers
Useful for delicate work such as removing seals and springs during faucet repairs.

Rib-joint pliers
Also known as slip-jaw pliers; open wide enough to remove drain traps.

Screwdrivers
Both slot *(top)* and Phillips *(bottom)* are common household tools essential for fixing leaking faucets and making other repairs.

Drain-and-trap auger
Also known as a snake; stretches 10' to 25' to remove deep drain blockages.

Toilet plunger
With its funnel cup, dislodges clogs by alternating pressure and suction.

Sink plunger
Flat face for drains.

Toilet auger
Works like a drain-and-trap auger to unclog toilets; 3' to 6' tool with a crank handle has a bent housing to get into bowl's passage.

GETTING STARTED IN PLUMBING

To do any job right, you need the proper tools and supplies—and plumbing is no exception. There are specific plumbing tools, as well as other standard tools which you may possess already. A number of supplies required for various plumbing tasks are listed below. Finally, if you'll be installing any new runs of pipe, make sure you have the correct type of supports *(below)* for the pipe you're using. CAUTION: Do not use one type of metal hanger to support another type of metal pipe or tube.

The tools shown opposite are frequently used in a wide range of plumbing repairs and improvements. Also, consult the following list of everyday tools to complete your plumbing toolkit:
- Saw (use a fine-tooth blade with 24 to 32 teeth per inch)
- Miter box (for square cuts in pipe and tube)
- Compass saw (used for cutting curves in wood, plywood, or gypsum wallboard; should have a blade with 7 or 8 teeth per inch)
- Saber saw (power saw used to cut curves in wood, metal, and plastic)
- Reciprocating saw with metal-cutting blade (for cutting cast-iron or galvanized steel pipes)
- Open-end wrenches ($3/8$ inch to $3/4$ inch to grip nut head from side)
- Hex wrenches (various sizes for driving setscrews)
- Locking pliers (adjustable width and jaw tension to clamp firmly; 8-inch model is best for plumbing work)
- Propane torch (used for sweat soldering copper tube and fittings)
- Knife (pocketknife or utility knife for deburring plastic pipe and tube)
- Putty knife
- Flashlight and/or worklight
- Drill (standard electric or cordless drill and an array of bits)
- Half-round file (has a round and a flat side for use on flat or concave surfaces)
- Claw hammer
- Cold chisel and ball-peen hammer (to cut metal pipe)
- Tape measure
- Carpenter's level
- Wire brush

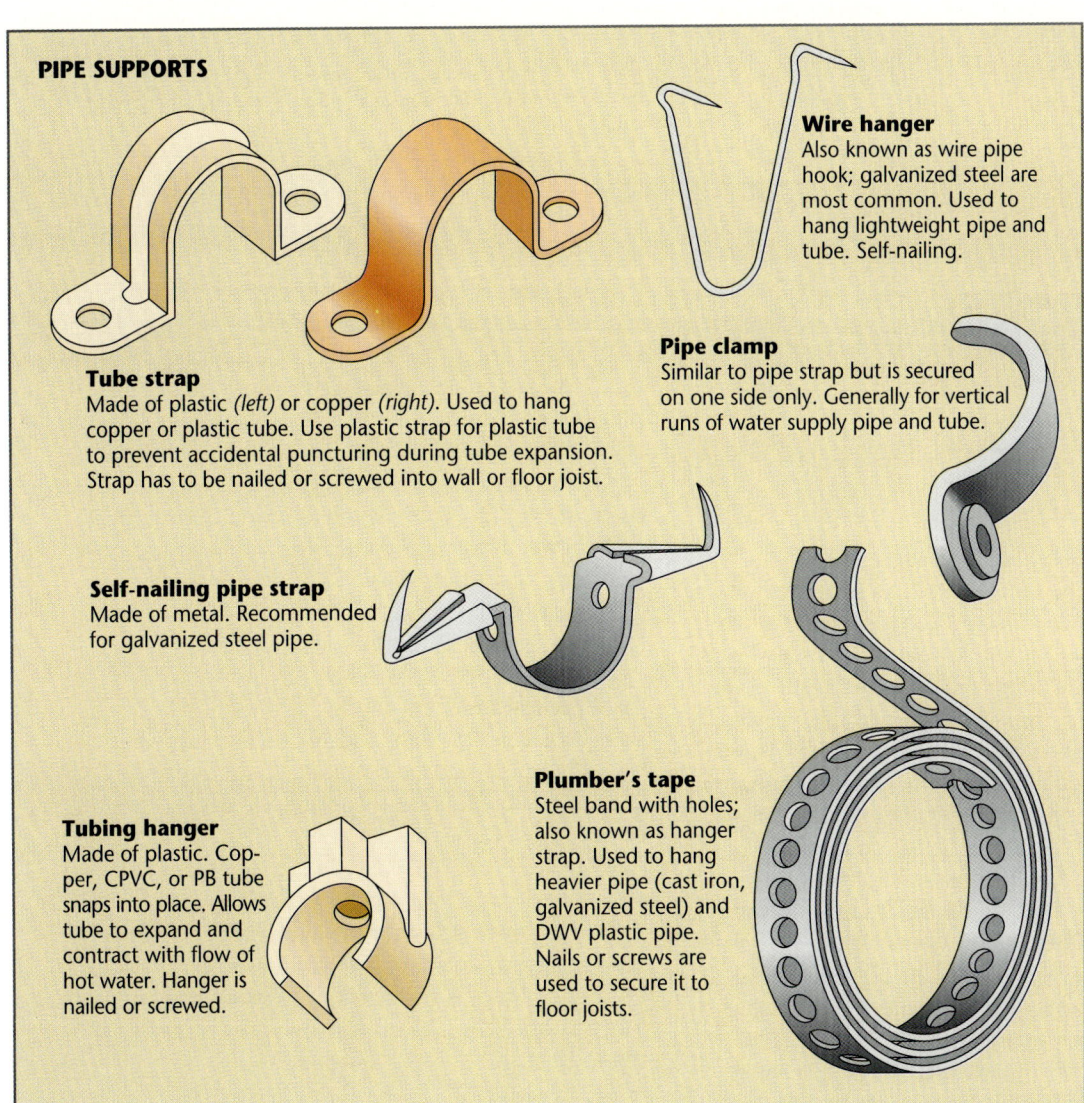

PIPE SUPPORTS

Tube strap
Made of plastic *(left)* or copper *(right)*. Used to hang copper or plastic tube. Use plastic strap for plastic tube to prevent accidental puncturing during tube expansion. Strap has to be nailed or screwed into wall or floor joist.

Self-nailing pipe strap
Made of metal. Recommended for galvanized steel pipe.

Tubing hanger
Made of plastic. Copper, CPVC, or PB tube snaps into place. Allows tube to expand and contract with flow of hot water. Hanger is nailed or screwed.

Wire hanger
Also known as wire pipe hook; galvanized steel are most common. Used to hang lightweight pipe and tube. Self-nailing.

Pipe clamp
Similar to pipe strap but is secured on one side only. Generally for vertical runs of water supply pipe and tube.

Plumber's tape
Steel band with holes; also known as hanger strap. Used to hang heavier pipe (cast iron, galvanized steel) and DWV plastic pipe. Nails or screws are used to secure it to floor joists.

PLUMBING SUPPLIES YOU MAY NEED

- Pipe-joint compound
- Plumber's putty
- Pipe-thread tape
- Flux and solder
- Penetrating oil
- For emergencies: bicycle inner tube, old hose, automotive hose clamps, faucet washers, wire coat hangers, nuts, bolts, and metal washers
- Antifreeze (to winterize drains in cold climates)
- Steel wool
- Brush
- Plumber's grease
- Silicone grease (for rubber parts)

GETTING STARTED IN PLUMBING

WORKING SAFELY

It's important to play it safe at all times. Wear appropriate safety gear, shut off the water before beginning any plumbing job, and always consult your local plumbing code. Wear safety goggles whenever you cut or hang pipe or tube, drive nails, work with power tools, do sweat soldering, or use drain cleaners. Leather work gloves protect hands from abrasions while cutting or threading pipe and tube.

Wear rubber gloves when cleaning out drain clogs or using drain cleaners. There may be septic material in the water, and the caustic chemicals used in drain cleaners can burn if they come into contact with skin. The work area should be well ventilated. Exercise common sense when using any hand or power tools. If you don't know how to use them, find out first.

Remember that water and electricity don't mix! Before installing any electrical appliances, such as a dishwasher *(page 88)*, consult an electrician. A professional should also be consulted before beginning work on any gas appliance.

Turning off the water

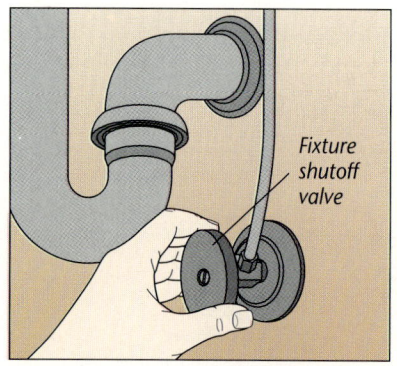

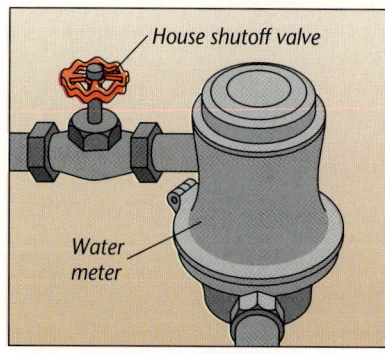

Using shutoff valves
The water supply can be turned off at the fixture shutoff valve and at the house shutoff valve. You'll usually find the fixture shutoff valve directly under the fixture or appliance at the point where the water supply connects to it *(above, top)*. If the fixture does not have a shutoff valve, one can easily be added, as shown on page 75.

Use the house shutoff valve *(above)* to turn off the water supply to the house. Know where this valve is and label and test it before any trouble arises. The house shutoff valve is usually a gate-type and is close to the point where the water supply enters the house, either inside or outside.

PLUMBING CODES AND PERMITS

Plumbing codes and permits establish standards for materials and workmanship to protect the health and safety of the community, and ensure that those standards are followed. Faulty plumbing can cause serious health and safety hazards such as contamination of drinking water, toxic gas backups, burst pipes, floods, and electrical shorts.

Just like a contractor, a do-it-yourself plumber must abide by all the rules and regulations of the codes. If your work violates a code, you run the risk of having to rip it all out and do it again.

There are six regional plumbing codes in print, but regulations regarding methods, materials, and design differ from one state, county, or municipality to the next. Local codes supersede the regional codes. A nationwide plumbing code is under consideration and may soon be in effect.

Codes are not laws, but regulations that carry the force of the law. This means that, if not adhered to, plumbing codes can be enforced through the police power of local government. A local plumbing inspector is responsible for seeing that the plumbing code is followed. Seek advice and follow any given.

Before you begin any work, be sure your plumbing plans conform to local codes and ordinances. Discuss your plans in detail with a local building inspector and be sure the methods and materials you're planning to use are acceptable. The inspector will tell you whether or not you need a plumbing or building permit. Projects that involve changes or additions to your plumbing system—specifically, to the pipes—usually require permits. You won't need a permit, though, for replacements—such as a new fixture or appliance—or for emergency repairs, as long as the work doesn't alter the plumbing system. When in doubt, be sure to check with the inspector.

WORKING WITH PLASTIC PIPE AND TUBE

Most homeowners who have worked with various types of pipe prefer to work with plastic. That's because plastic piping is lightweight, inexpensive, and easy to cut and fit. Unlike metal pipe, plastic is also self-insulating and resistant to damage from chemicals and electrolytic corrosion. In addition, plastic's smooth interior surface provides less flow resistance than metal.

All the major plumbing codes and the Federal Housing Administration accept plastic pipe, even though a few local codes may not. More than a million homes have been plumbed with it. To make sure the pipe you are getting is accepted, look for the American Society for Testing and Materials "ASTM" designation printed on the pipe. If the pipe is to carry potable water, it also should have the National Sanitation Foundation's "NSF" seal. Avoid products without these designations.

Plastic piping is sized nominally, according to inside diameter, as being either pipe-sized or tube-sized. Pipe-sized means following iron-pipe sizing; tube-sized means following copper-water-tube sizing. All plastic drain-waste-vent piping is pipe-sized, as is plastic piping for outdoor cold-water-only use. And all plastic water supply piping is tube-sized. Tubes are sometimes designated by their outside diameter. In either case, outside diameter (O.D.) is stated.

TYPES OF PLASTIC PIPE AND TUBE

Plastic is either flexible or rigid, with fittings that turn, branch off, reduce, and adapt to other types of piping. Rigid types of plastic pipe are most often joined to their fittings by solvent welding, while fittings for flexible plastic pipe are joined mechanically (by hand).

PLASTIC PIPE AND TUBE

Type	Characteristics	Cutting tools required	Joining method	Support required
PB (flexible) (tube-sized)	Grey or off-white; used for hot and cold water supply systems or riser tubes, indoors or outdoors. Available in 5' straight or 25' to 100' coils; 1/4" to 3/4" nominal inside diameter.	Plastic tubing cutter/sharp knife/miter box and fine-toothed saw	Mechanical O-ring	Plastic strap hangers every 32"
PE (flexible) (pipe-sized)	Black; used for irrigation systems, water wells, or water service entrances, outdoors only. Available in 25' to 400' coils; 1/2" to 2" inside diameter in various pressure ratings.	Fine-toothed saw/sharp knife	Mechanical insert fittings/worm-drive band clamps	Rock-free soil
CPVC (rigid) (tube-sized)	Grey or off-white; used for hot and cold water supply systems indoors, outdoors, or underground. Available in 10' straight lengths; 1/2" or 3/4" nominal diameter.	Plastic tubing cutter/miter box and fine-toothed saw	Solvent welding/mechanical O-ring	Plastic strap hangers every 32" (horizontal); every story or 10' (vertical)
PVC (rigid) (pipe-sized)	White, or, for DWV use, white and off-white; used for cold water supply systems, irrigation systems, DWV systems, sewer piping, or drain traps, indoors, outdoors, or underground. Available in 10' to 20' lengths; 1/2" to 2" inside diameter; DWV 10' to 20' lengths; 1 1/2" to 4" inside diameter.	Plastic tubing cutter/miter box and fine-toothed saw	Solvent welding	DWV: Plumber's tape every 4' (horizontal); every story or 10' (vertical) Supply: Rock-free soil
ABS (rigid) (pipe-sized)	Black; used for DWV systems or drain traps, indoors or outdoors. Available in 10' to 20' lengths; 1 1/2" to 4" inside diameter.	Plastic tubing cutter/miter box and fine-toothed saw	Solvent welding	Same as PVC DWV *(above)*
RS (rigid) (pipe-sized)	White or off-white; used for outdoor sewer systems, septic systems 5' from house, storm drains. Available 10' to 20' long; 4" nominal diameter.	Fine-toothed saw	Solvent welding	Rock-free soil
PP (rigid) (tube-sized)	Off-white; indoor use only. Comes as drainage fixture traps and drain tubes.	No tools required	Mechanical drainage tube slip couplings	No supports required

GETTING STARTED IN PLUMBING

Flexible plastic includes PB (polybutylene) tubing and PE (polyethylene) pipe (see the chart on page 11). PB is used for long, joint-free runs (such as water service entrances) or for remodeling work where the tubes can be snaked between walls, floors, and ceilings. PB riser tubes connect the water supply system and faucets and toilets in cramped spaces. They are also used to connect fixture shutoff valves.

There are five types of rigid plastic piping: CPVC (chlorinated polyvinyl chloride); PVC (polyvinyl chloride) and ABS (acrylonitrile-butadiene-styrene), used for DWV systems; RS (rubber styrene) for sewer and drainage pipe; and heat- and chemical-resistant PP (polypropylene), used in fixture traps and drain tubes.

For DWV use, one type of PVC piping, called Schedule 30, has slightly thinner walls so that the pipe and its fittings will fit a standard 2x4 stud wall without special treatment. Check your code before using. Both PVC and ABS are less expensive, lighter in weight, and easier to connect and hang than sections of cast-iron pipe *(page 25)*. For these reasons, plastic pipe is a common choice for extending a cast-iron system, and even for replacing a leaking cast-iron pipe.

PVC is a better choice than ABS because it is less susceptible to mechanical and chemical damage and has a slightly greater variety of fittings available (fittings for PVC and ABS are not readily interchangeable). PVC, while not heat-resistant, is also used for traps and drains. Don't store plastic pipe in direct sunlight because the ultraviolet rays can damage it.

PRESSURE PRECAUTIONS

Plastic water supply pipes and tubes have pressure and temperature ratings, designated on the sidewalls or as part of their ASTM rating. These should not be exceeded, although this is seldom a problem. For example,

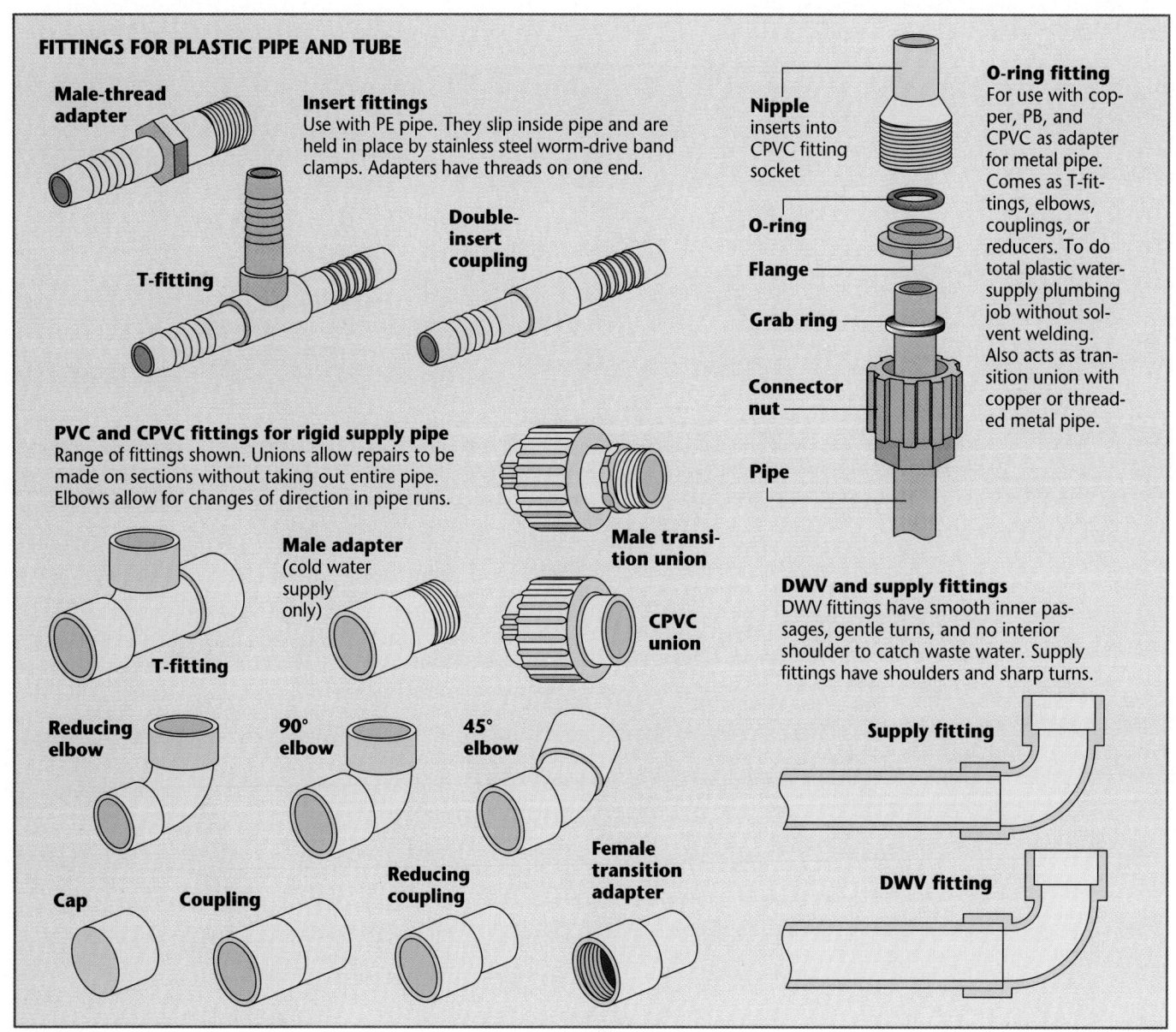

both rigid CPVC and flexible PB are rated to withstand water at a temperature of 180°F continuously at 100 psi pressure, which exceeds what would be found in a typical house.

Because the pressure ratings for plastic tube are lower than those for metal, plastic systems are less able to withstand line surges (sudden changes in water pressure in the system) caused by water hammer, for example. To prevent problems, it's important to install water-hammer arresters at all fixtures and water-using appliances (except toilets). This is a good plan for any water supply system, plastic or metal.

FITTINGS

Plastic pipe and tube can be joined with a number of fittings. For example, PE pipe is joined by insert fittings, as shown opposite. The pipe needn't be cut off squarely before the fittings are inserted. If you need to make changes after insertion, usually the joint can be pulled apart by hand once the clamps have been loosened. If not, simply pour hot water over the ends of the PE pipe to soften it. Then pull.

The easiest, least expensive way to connect PB and CPVC is with O-ring fittings. Made of CPVC, they also work with copper tubing. Using an O-ring T-fitting, it's possible to join a rigid CPVC tube, a flexible or rigid copper tube, and a flexible PB tube all in one fitting.

A neoprene rubber O-ring held inside the fitting's body by a flange makes a watertight seal between the fitting and tube. A stainless steel grab ring locks around the outside of the tube. The tube can be inserted past the grab ring but cannot be pulled out—or blown out—by water pressure. A hand-tightened connector nut holds the assembly together. No tools, clamps, or solvent cement are needed to make a connection. Push the tube into the fitting until it bottoms out on the inside shoulder. To make sure the tube is inserted all the way, mark the tube 1½ inches from the end. When the mark meets the connector nut, you are assured of full insertion. It helps insertion to chamfer the tube end with a knife and coat it with hand soap. Tighten the connector nut. While the joint will not come apart, it may be taken apart and put back together.

Most PVC and CPVC fittings push onto the ends of pipes and are joined by permanent solvent cement. Transition fittings—those that let you link plastic pipe to piping of a different material—often have threads on one end. Threaded fittings, called unions, make it easy for you to change or extend supply tube by unscrewing a length of tube in the middle of a run. Reducer fittings allow you to link tubes of different diameters.

DWV fittings and supply fittings differ from each other in that DWV fittings have smooth bends and no interior shoulder that could catch waste. You use a different method to assemble DWV fittings, depending on whether you're joining plastic to plastic, plastic to lengths of hubless cast iron, or plastic to the bell of bell and spigot (hub) cast iron. Fittings for joining plastic to plastic and plastic to lengths of hubless cast iron *(page 16)* include reducer fittings that allow you to connect pipes of different diameters; always direct the DWV flow from the lesser to the greater pipe diameter.

Plastic DWV pipe is joined by solvent welding *(page 15)*. See page 16 for detailed instructions on assembling DWV fittings.

Supply fittings are smaller than DWV fittings and may have sharper turns with restricting shoulders. They are designed to work under pressurized conditions, unlike DWV fittings which must work with the flow of gravity.

TAPPING INTO PLASTIC PIPE

Since the widespread use of plastic tubing indoors is a recent trend, you'll encounter either copper tube *(page 17)* or galvanized steel pipe *(page 22)* when you penetrate a wall *(page 68)*. If you do discover a run of plastic, it's likely to be CPVC using solvent-welded joints. To tap in, cut out a short section of pipe and solvent weld a CPVC T-fitting in place of the cutout. Or install a push-in, hand-tightened O-ring T without solvent welding.

Before cutting, have a pail or absorbent cloth in place to catch any spills. CAUTION: Turn off the water supply at the house shutoff valve *(page 10)*. Then drain the pipes by opening a faucet at the low end. Use a plastic tube cutter or any saw with a fine-toothed blade (24 or 32 teeth per inch, or tpi). While you saw, brace the tubing to prevent excess motion that might strain the fittings.

ASK A PRO

CAN I JOIN PLASTIC TO METAL?

Plastic and metal expand and contract at differing rates, so whenever plastic tubing carrying pressurized hot water is joined to metal piping, a fitting called a transition must be used to prevent leaks. These connections occur at water heaters, tub and shower valves, and anywhere else threaded connections are needed. Because heat may be conducted to both sides, a valve's cold side is also fitted with a transition union. Simple male-thread adapters may be used to make non-pressurized connections to female threads, at a shower riser, and for cold-water-only connections.

MEASURING, CUTTING, AND HANGING PLASTIC PIPE AND TUBE

Before you cut any pipe, make exact measurements. Rigid pipes won't give if they're too long or too short. It's not a problem for flexible tube due to its flexibility.

To measure pipe and tube, determine the face-to-face distance between new fittings, then add the distance the pipe will extend into the fittings *(right)*. If you use two short pieces of pipe connected by a union, don't forget to count the union as a third fitting.

The distance the new pipe will extend into the fittings, called makeup, depends on the type of fittings. In push-on fittings, pipe ends extend all the way to the shoulder once the solvent cement is brushed on; in threaded fittings, they don't go quite so far. (Solvent welding for rigid supply pipe and tube is shown on the opposite page; see page 16 for joining DWV fittings.)

Cut flexible or rigid plastic pipe and tube with a plastic tubing cutter. You can also use a pipe cutter fitted with a special plastic-cutting wheel or a miter box and fine-toothed saw with 24 or 32 tpi *(opposite)*.

Use the same method to measure and cut plastic DWV pipe. To remove a section of cast-iron DWV pipe to make room for plastic pipe, see page 26. To replace a section of plastic DWV pipe, see pages 15 to 16.

When hanging pipe, don't bind runs of rigid plastic supply pipe; use plastic hangers that hold it snugly yet allow it to move with expansion and contraction. Support vertical runs of pipe at every story. Pipe hang-

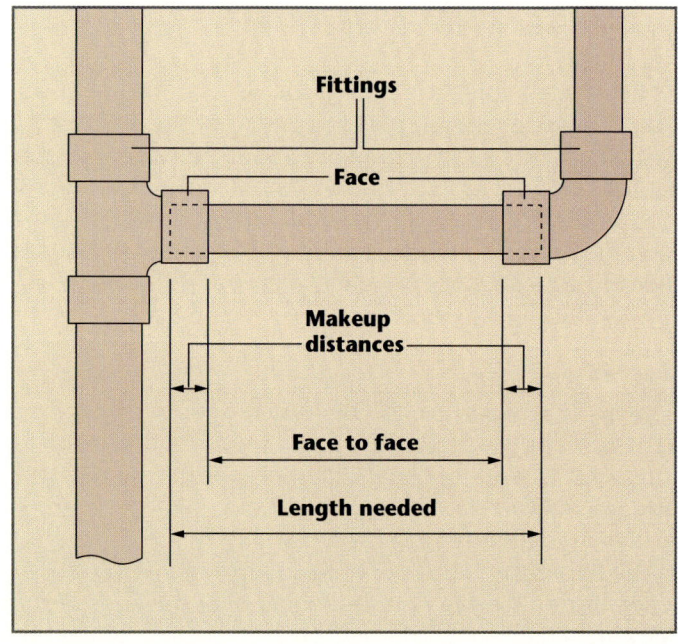

ers are on page 9; see the chart on page 11 for proper support distances. Wear safety goggles when cutting pipe or tube overhead.

For DWV pipe, provide 1/4 inch of downward slope toward the house drain for every foot of horizontal run *(page 70)*.

Cutting flexible pipe or tube

TOOLKIT
- Sharp knife or plastic tubing cutter

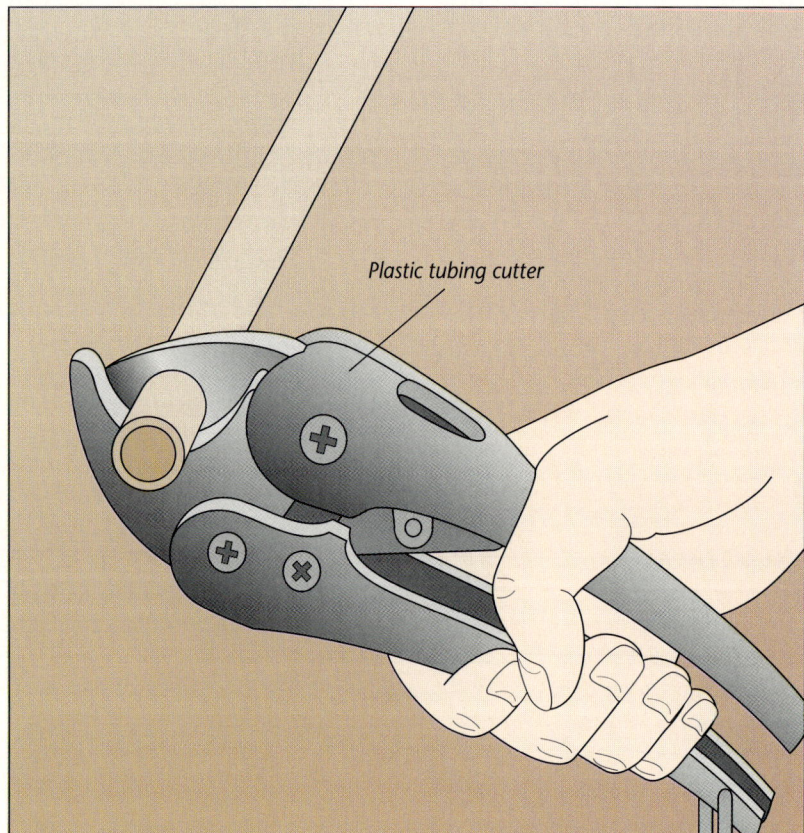

Using a plastic tubing cutter
Cut flexible or rigid plastic tube with a sharp knife or a plastic tubing cutter, with a cutting diameter of up to 1". To use the cutter, place the tube between the open jaws and close the handles to press down on the tube *(left)*, continuing until you cut through. Don't cut at an angle, and, if you're cutting flexible tube, leave a long enough piece of the tube to be able to make smooth bends.

14　GETTING STARTED IN PLUMBING

Cutting rigid pipe and tube

TOOLKIT
- Plastic-tubing cutter or pipe cutter or miter box
- Fine-toothed saw

Using a saw
You can cut rigid plastic pipe and tube with either a plastic-tubing cutter, a pipe cutter, or a miter box and a fine-toothed saw, such as a backsaw—preferably one with 24 to 32 teeth per inch *(right)*. If you use a saw on an installed pipe, brace the pipe to prevent excess motion that could affect the squareness of the cut.

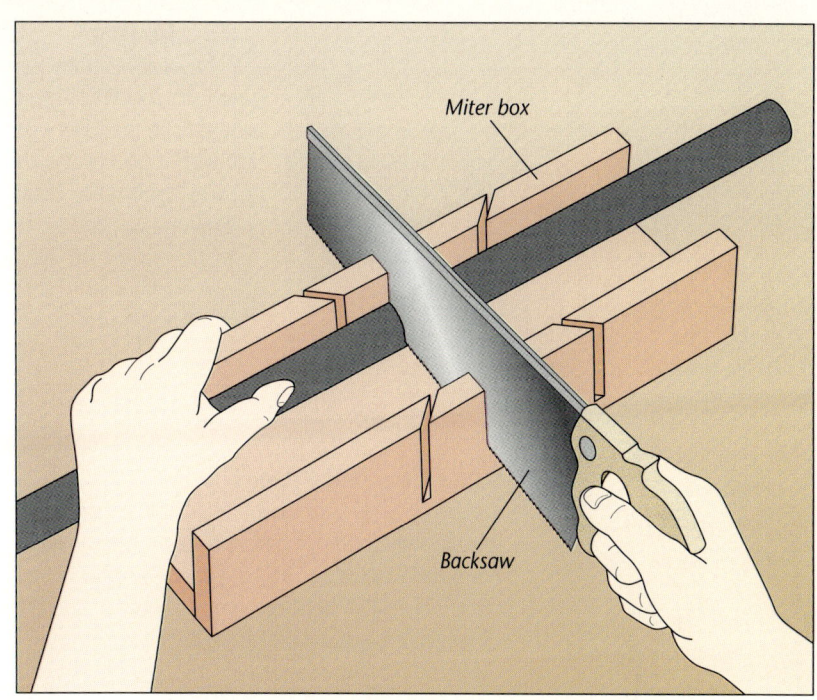

Solvent welding rigid pipe and tube

TOOLKIT
- Knife or deburring tool
- Soft brush

1 ▶ Preparation and test-fitting
Cut the pipe or tube squarely, then use a knife or a deburring tool to remove any burrs inside and outside the pipe end *(right)*. Inspect the end for cracks, gouges, and deep abrasions. Cut it again if necessary. Then test-fit the pipe in the fitting. It should enter the fitting, yet stop part-way in. When the assembly is inverted, the fitting shouldn't fall off. A successful solvent weld can't be made if the fit is too tight or too loose. It's a good idea to mark the pipe and the fitting for alignment beforehand and line up the two marks when cementing. The pipe won't slide into the fitting completely until the cement (which acts as a lubricant) is applied; make your marks long enough to take this into account.

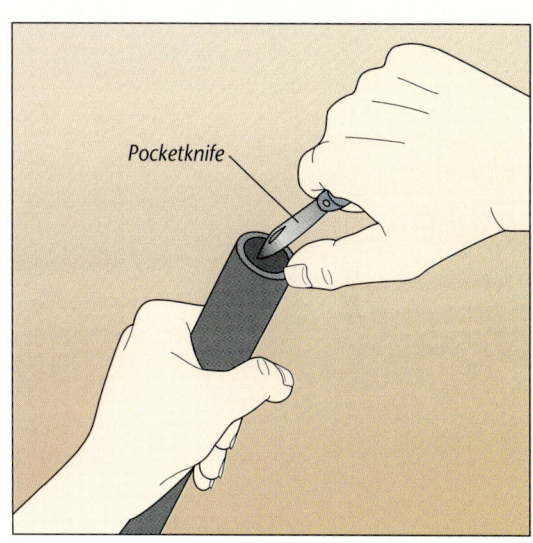

2 ▶ Applying solvent
Get the right type of solvent cement for the kind of plastic you're joining (an all-purpose solvent cement is also available). Usually the container will have an applicator brush. If a brush isn't included, use a soft brush—for example, a 1/4" brush for 1/2" pipe, or a 3/8" brush for 3/4" pipe. Work in a well-ventilated area, avoid breathing fumes, and keep lighted matches and cigarettes away from the flammable solvent cement.

Following the manufacturer's instructions, apply solvent cement liberally to the pipe, then more lightly to the fitting socket *(left)*.

GETTING STARTED IN PLUMBING

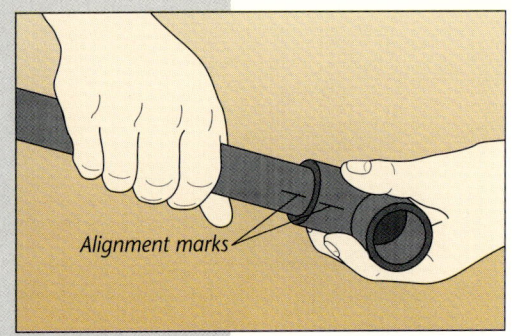
Alignment marks

> **3 Joining**
> Immediately join the pipe and fitting with a slight twist, bringing the fitting into correct alignment *(left)*. Hold for a few seconds while the cement sets. Inspect the joint between pipe and fitting. There should be a fillet of solvent all around. If you make an error, cut off the fitting and replace it with another, joining it with two couplings and two short lengths of pipe.
>
> Solvent-welded joints can be handled gently within a minute, but wait at least two hours before pressurizing with water, and even longer under cold, damp conditions. If the temperature is below 40°F, use a special low-temperature solvent cement. Then you can turn on the water and inspect for leaks.

Attaching screw-on fittings to rigid pipe

Using threaded fittings
Once in a while, you'll encounter rigid plastic pipe whose ends or fittings have exterior threads (most common with the 1" to 12" precut lengths of thick-walled PVC pipe known as nipples). Such pipe requires special plastic fittings with interior threads. To make a watertight seal, first wrap pipe-thread tape 1 1/2 turns clockwise around the threads of the pipe, pulling the tape so tight that the threads show through. Then, screw on the fitting. Plastic threads are not tightened as tightly as metal threads. For a proper fit, screw the fitting in by hand, then give it one turn with a wrench beyond hand-tight. Don't overtighten. Threads should still be showing when you finish.

Assembling DWV fittings

TOOLKIT
For joining to bell-and-spigot cast iron:
• Torch
• Hammer
• Putty knife
For joining to no-hub cast iron:
• Cast-iron snap cutter
• Plastic tubing cutter or miter box and hacksaw

Joining plastic to plastic
Use solvent cement to join the pipe to its fittings. When you're adding DWV pipe, make sure you securely brace the cut pipe already in the wall or floor with plumber's tape before adding new pieces.

If there's not enough clearance between the free end of a new fitting and the pipe already installed to slip the fitting over the pipe, cut the pipe back and make the connection either with short lengths of pipe and slip couplings or no-hub connectors. Slip couplings are like normal couplings, but without center shoulders so they can slide all the way onto the pipe ends and be out of the way.

Joining plastic to bell-and-spigot cast iron
To remove a length of cast-iron pipe from a bell, cut out a piece *(page 26)*, leaving a short length in the bell. Remove the short length by working it back and forth while you melt the lead in the joint with a torch. Join the plastic to the bell by packing the joint with oakum (stranded hemp fiber), then hammer in cold lead wool to make the joint watertight, or pack a plastic, puttylike lead substitute into the joint over the oakum with a putty knife. These materials are available at plumbing supply houses. The oakum seals the joint, so pack it tightly; the plastic only keeps the oakum in.

Joining plastic to no-hub cast iron
Cut out the section for replacement or a new fitting as described on page 26. To replace a damaged cast-iron run with plastic, cut the plastic pipe to the exact length of the opening, place the new run in position (you may need a helping hand), and make the connections with no-hub couplings.

To extend new plastic pipe from a cast-iron system, you have a choice of methods. You can add a no-hub sanitary T-fitting with couplings *(page 70)* and run plastic from that point; or, you can install a plastic sanitary T-fitting with spacers (stubs of plastic pipe) and no-hub couplings *(left)*.

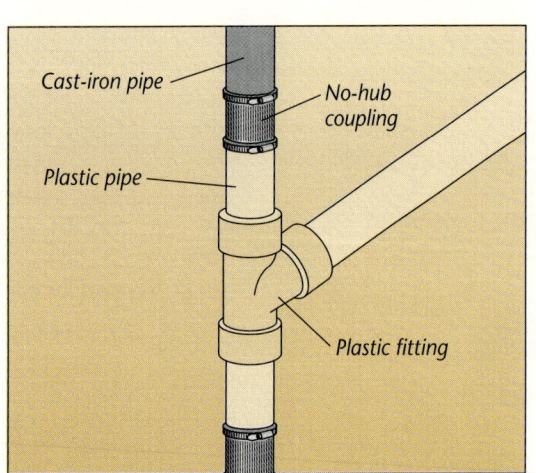
Cast-iron pipe — *No-hub coupling*
Plastic pipe
Plastic fitting

16   GETTING STARTED IN PLUMBING

WORKING WITH COPPER TUBE

Copper tube is lightweight, fairly easy to join (by sweat soldering or with flare, compression, or union fittings), resistant to corrosion, and rugged. Its smooth interior surface allows water to flow easily.

Two kinds of copper tube—hard- and soft-temper—are used in supply systems to carry fresh water. Another type of copper tube—corrugated supply—is used as flexible tubing to link hard or soft tube to fixtures. Copper pipe, with a larger diameter, is used in drain-waste-vent (DWV) systems.

Hard-temper copper: This is sold in lengths of 20 feet or less. Because it can't be bent without crimping, it must be cut and joined with fittings whenever there is a change in direction. It comes in three thicknesses: K (thick wall), L (medium wall), and M (thin wall); M is usually adequate for home plumbing. Nominal diameters range from 1/4 to 1 inch and larger; actual diameters are greater. For tube-sized piping, the outside diameters of the same sizes are equal but the inside diameters vary; those with thicker walls are smaller on the inside.

Soft-temper copper: This is sold in 20-, 60-, and 100-foot coils. More costly than hard-temper supply tube, its offers the advantage of not needing as many fittings, since it can be bent without crimping. Soft-temper copper is available in K and L thicknesses; L is adequate for above-ground plumbing. Nominal diameters range from 1/4 to 1 inch; actual diameters are slightly more.

Copper DWV pipe: This is usually sold in 20-foot lengths and in nominal diameters of 1 1/2, 2, and 3 inches. (Copper DWV pipe with nominal diameters larger than 2 inches is too expensive to be readily available.)

Flexible tubing: Used for linking supply tube to fixtures, this corrugated, smooth, or chrome-plated copper comes in short lengths. It can conform to tighter curves than soft copper tube, and has a nominal diameter of 3/8 or 1/2 inch. It often comes in kit form; follow the manufacturer's instructions.

CUTTING AND JOINING COPPER TUBE

Before beginning work, turn off the water supply at the house shutoff valve *(page 10)*. Drain the pipes by opening a faucet at the low end. To replace leaking copper tube, or to extend supply pipes, you'll need to know some basic removal, measuring, and cutting techniques *(pages 18 to 19)*. These apply equally to hard- and soft-temper supply tube. DWV pipe is measured and cut in the same manner, but it usually needs to be braced like cast-iron DWV pipe *(page 70)* before removal.

Different joining methods require different fittings *(page 18)*. Sweat soldering is the best way to keep copper tube (hard or soft) together. Hard-temper supply tube can be joined with compression fittings, soft-temper supply tube with compression or flare fittings. Copper's softness means it can't be successfully threaded.

If you think you'll need to take apart a run of copper tube (to replace a water heater, for instance), without unsoldering or cutting the run, fit two short lengths of the pipe together with a union *(page 21)*.

Reducer fittings allow you to link pipes of different diameters; transition fittings let you join copper tube with plastic tube or galvanized steel pipe. If you link copper with galvanized steel, though, you must use special dielectric fittings, to prevent electrolytic corrosion *(page 22)*.

Soldered joints are often called sweat joints, and are made with copper fittings that have smooth interiors. To sweat solder a joint *(page 20)*, you'll need a small propane torch, some 00 steel wool, very fine sandpaper or emery cloth, a can of soldering flux, and some lead-free plumber's solder. CAUTION: Use only solder that's labeled lead-free in making joints in any potable water supply or DWV system. Wear safety goggles and comfortable work gloves while working.

A flared joint *(page 21)* is only for soft copper tube. Because it tends to weaken the end of the pipe, use a flared joint only if you can't solder and can't find the right compression fitting.

Compression fittings *(page 21)* work equally well on hard- and soft-temper copper tube, which gives them an advantage over flared fittings when creating nonsoldered joints. A further advantage is that, unlike flare fittings, compression fittings don't require a special tool during assembly.

Union joints *(page 21)* are available only as straight couplings. They only link tubes of the same diameter and are composed of three elements like compression fittings. Unlike compression fittings, the three elements of union joints allow you to install or easily remove a union without having to twist the tube itself.

Once lengths of tube are joined, support the run every 6 to 8 feet with one of the hangers shown on page 9. Unless the hanger is made of copper, insulate the tube from the hanger so that electrolysis can't occur; do this by wrapping the tube with electrician's tape where the hanger would touch it.

ASK A PRO

HOW DO I HANDLE COPPER?
Whether the tube is "hard" or "soft," all copper is a soft metal, so you'll want to be careful not to damage it as you work. Don't use wrenches and vises which could crush the metal.

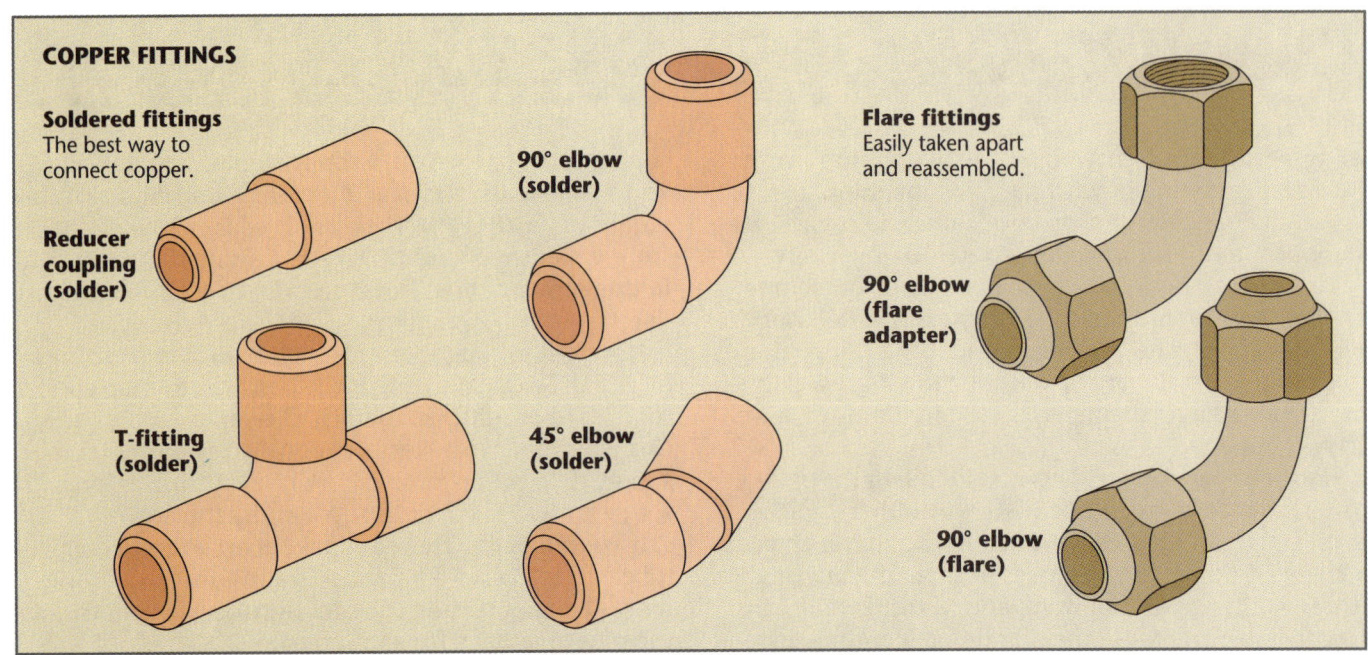

Removing copper tube

TOOLKIT
- Fine-toothed hacksaw (24 to 32 tpi)

OR
- Butane or propane torch

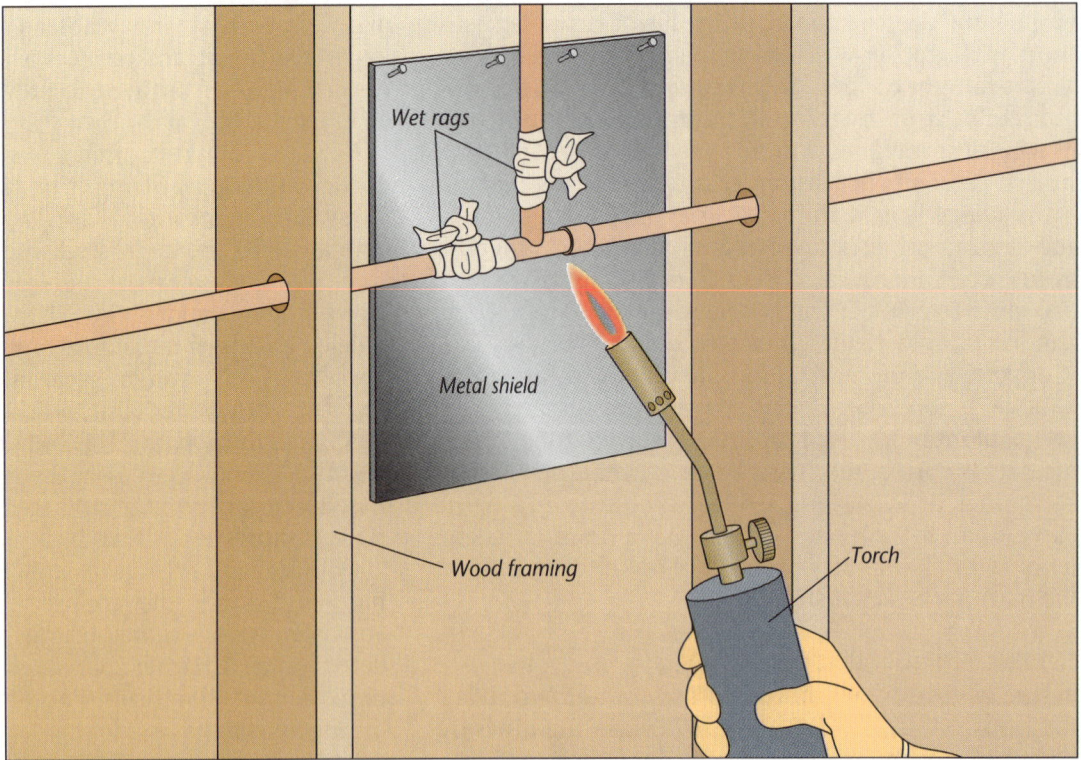

Using a saw or torch

Shut off the water *(page 10)* and drain the tube by opening a faucet at a lower end. If the copper is joined with compression, flare, or union fittings, uncouple them by unscrewing the fitting. For soldered joints, you'll need to cut the tube with a saw, or, if you can pull the ends of the tube free once the solder joint is loosened, melt the solder with a butane or propane torch.

To use a saw, brace the run to prevent excess motion that would strain joints and cause severed ends to sag. If the run is not supported, wrap plumber's tape once around the tube every 3'; pull the tape taut and nail it to nearby joists or studs.

To use a torch, shield flammable material with a piece of metal and wrap wet rags around the tube at joints you wish to leave intact *(above)*. Have a fire extinguisher on hand; you don't want to start a house fire while the water is turned off.

CAUTION: Be sure to wear safety goggles while sawing or using a torch.

Measuring copper tube

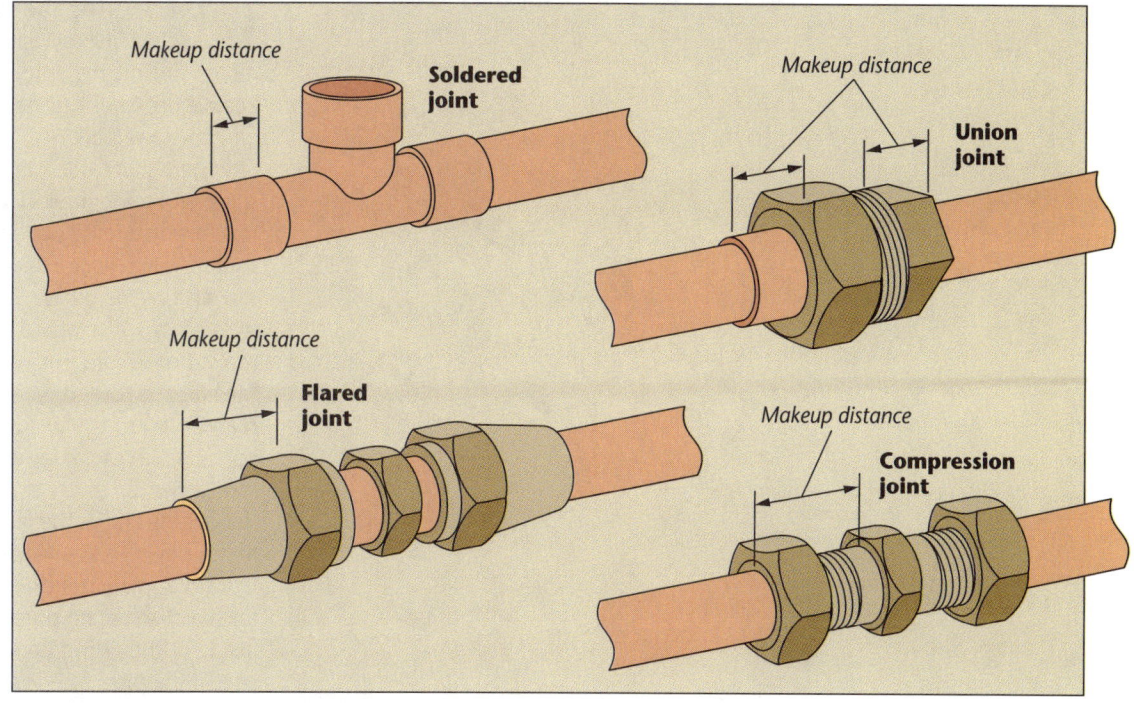

Calculating the distance
To determine how much new copper tube you need, measure the distance between new fittings, then add the distance the tube will extend into the fittings. Makeup distances vary for different types of joints *(above)*.

Cutting copper tube

TOOLKIT
- Pipe cutter or fine-toothed hacksaw
- Round file or pipe reamer

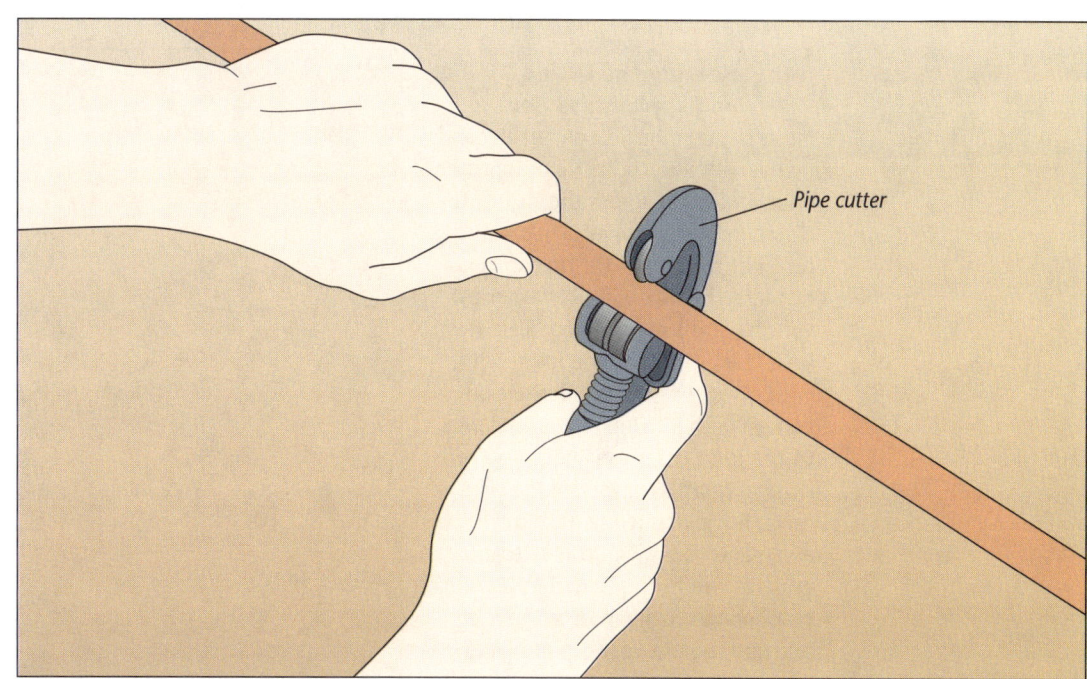

Using a pipe cutter
Cut new lengths of copper tube with a pipe cutter designed for copper tube. To use the cutter, twist the knob until the cutter wheel makes contact with the surface. Rotate the cutter around the tube *(above)*, tightening after each revolution, until the tube snaps in two. You can also cut copper tube with a fine-toothed hacksaw or mini-hacksaw (24 to 32 teeth per inch), but it's more difficult to make a straight, clean cut with a saw than with a pipe cutter.

After you've cut the tube, clean off inside burrs with a round file or with the retractable reamer often found on the pipe cutter. Then file or sand off outside burrs.

GETTING STARTED IN PLUMBING

Making a soldered joint

TOOLKIT
- Round file or pipe reamer
- Butane or propane torch

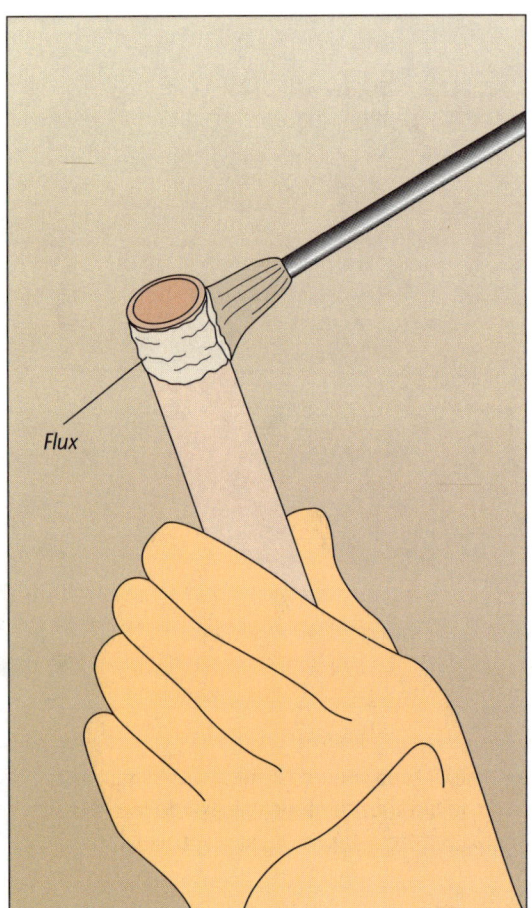

Flux

1. Preparing the fitting and applying flux

CAUTION: Wear safety goggles and work gloves during the soldering process. Be sure to keep a fire extinguisher on hand as the water will be turned off.

If there's any water in the pipes, it will hinder a successful soldering job. Dry the pipes as much as possible by turning off the water supply at the house shutoff valve *(page 10)* and opening a faucet at the low end of the pipes. Stuff the ends of the pipe with plain white bread to absorb any remaining moisture. Left in, the bread will disintegrate once the water is turned on. Using a round file or pipe reamer, ream the inside of the tube. File off any burrs on the outer edge with a flat file. Use steel wool, sandpaper, or emery cloth to polish the last inch of the outside end of the pipe and the inside end of the fitting down to the shoulder until they are shiny.

With a small, stiff brush, apply flux that's made for sweat-soldering purposes around the polished inside of the fitting and around the polished outside of the pipe end *(left)*. Avoid getting any flux on your hands, as the chemicals in it can be damaging. It's best to wear work gloves during the entire sweat-soldering process, as they can prevent burns, as well. Place the fitting on the end of the pipe. Turn the pipe or the fitting back and forth once or twice to spread the flux evenly.

2. Heating the fitting and applying solder

Next, position the fitting correctly and heat it with a torch, moving the flame back and forth across the fitting to distribute the heat evenly. It's important not to get the fitting too hot because the flux will burn—simply vanish—if it's overheated. Test the heat level this way: the joint is hot enough when solder will melt on contact with it. Touch the solder wire to the joint occasionally as you're heating *(right)*. The instant the wire melts, the joint is ready. Take the torch away and touch the solder to the edge of the fitting; capillary action pulls molten solder in between the fitting and the piping. Keep applying until a line of molten solder shows all the way around the fitting. After the joint cools, wipe off the surplus flux with a damp rag. Keep your hands away from the joint—the pipe can get quite hot as far as 1' or 2' on either side of the joint. Be careful not to bump or move the newly soldered joint until the solder hardens.

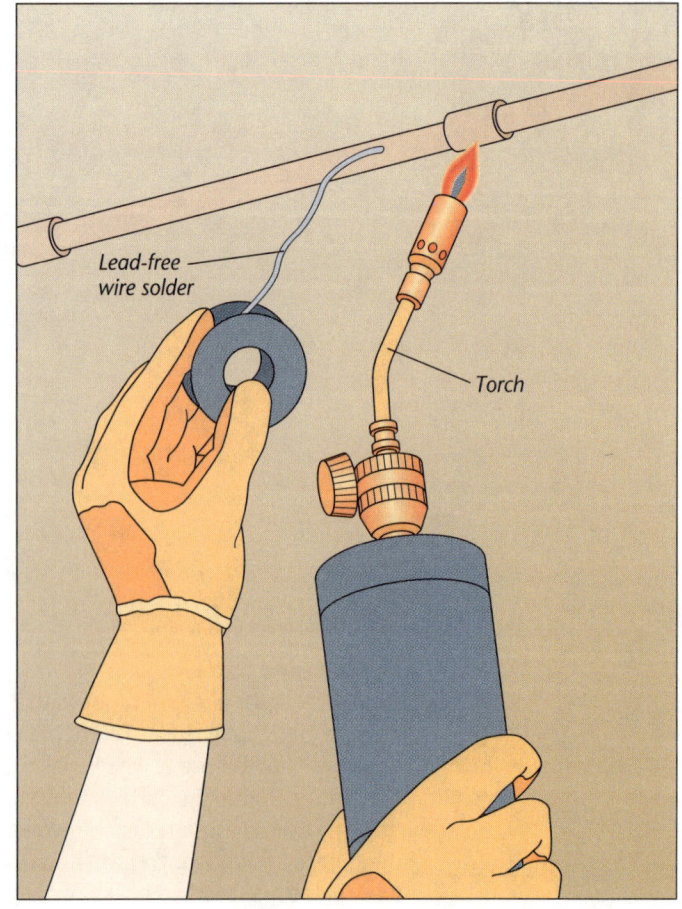

Lead-free wire solder

Torch

Making a flared joint

TOOLKIT
- Flaring tool
- 2 adjustable or open-end wrenches

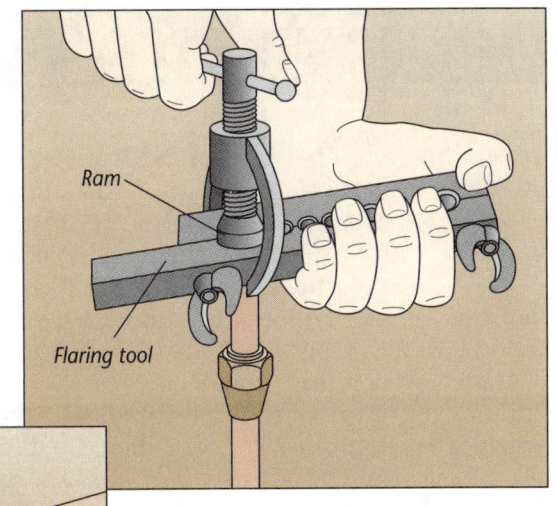

1. Flaring
To make a flared joint, slide the flare nut over the end of the tube, tapered end facing away from the end of the tube *(inset)*. Clamp the end of the tube into a flaring tool and screw the ram down hard into the end of the tube *(left)*. Remove the flared tube from the flaring tool.

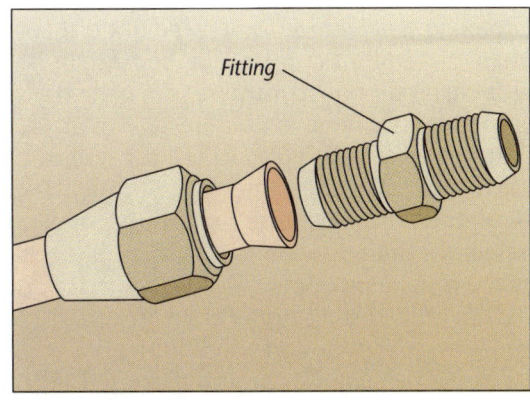

2. Attaching the nut
Press the tapered end of the body of the fitting into the flared end of the tube, and screw the nut onto the body of the fitting *(right)*. Use two wrenches to tighten.

Making a compression joint

TOOLKIT
- 2 adjustable or open-end wrenches

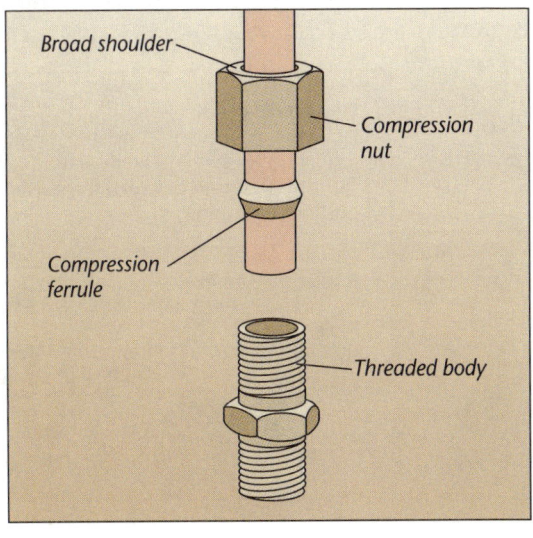

Using two wrenches
To install a compression fitting, slide the compression nut over the end of the tube, with the broad shoulder facing away from the tube's end. Then slip on the compression ferrule, as shown *(left)*. Push the threaded body of the fitting against the end of the tube, and screw the nut onto the body of the fitting. Tighten the nut with two open-end or adjustable wrenches, one on the nut and one on the body of the fitting. This compresses the ferrule tightly around the end of the tube and makes a watertight seal. Once compressed, the ferrule can't be removed.

Making a union joint

TOOLKIT
- Butane or propane torch
- 2 adjustable or open-end wrenches

Sweat-soldering and using two wrenches
To make a union, sweat solder the male shoulder onto one tube, then slip the nut onto the other tube. Sweat-solder the female shoulder onto the end of the second tube. Bring the male and female shoulders together, then slide the nut over the female shoulder, and screw the nut onto the male shoulder *(right)*. To tighten, use two wrenches—one holds the male shoulder, the other turns the nut.

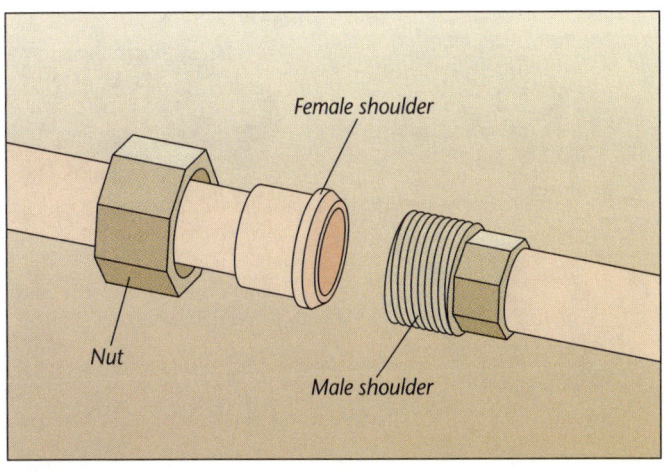

GETTING STARTED IN PLUMBING

WORKING WITH GALVANIZED STEEL PIPE

If your house is more than 20 years old, chances are you have galvanized steel supply pipe; you may even have galvanized steel drain-waste-vent (DWV) pipe, although cast iron *(page 25)*, or a combination of the two, is more common.

Galvanized pipe and its fittings are coated inside and out with zinc to resist corrosion. Despite this additional protection, galvanized steel pipe not only corrodes faster than cast iron or copper, but, because of its rough interior surface, collects mineral deposits that, over time, impede water flow.

It's common to replace a leaking length of galvanized steel pipe with the same type of pipe. It requires less equipment and expense than using copper, plastic, or cast iron. But when extending a supply system of galvanized pipe, use copper *(page 17)*, or plastic *(page 11)*.

Galvanized steel pipe comes in nominal diameters of $1/4$ to $2 1/2$ inches and in lengths of 10 and 21 feet, or it can be custom-cut and threaded. Also available are short, threaded pieces—called nipples—in $1/2$-inch increments from $1 1/2$ to 6 inches long, and then in 1-inch increments up to 12 inches long (the diameters match the pipes).

Fittings are connected to pipe by means of tapered pipe threads. Galvanized steel pipe is sold threaded on both ends. When you have pipe cut at the store, you might be able to have its new ends threaded there. If not, you can rent pipe-threading tools *(page 24)*.

Many fittings *(below)* are available for joining galvanized steel with the same type of pipe, or with copper or plastic. If you cut into galvanized pipe, you'll have to reconnect the ends with a union—a special fitting allowing you to join two threaded pipes without having to turn them.

Although all galvanized steel pipe is measured and cut in the same way, there are different removal methods. The facing page shows how supply pipe is removed. Galvanized steel DWV pipe uses the same method as cast-iron pipe *(page 26)*. Support horizontal runs of new pipe every 6 to 8 feet, and vertical runs every 8 to 10 feet; you can select from the different types of hangers shown on page 9.

Prevent cutting your fingers on newly cut pipe threads and injuring your hands while tightening, by wearing gloves. Soft leather work gloves are the most comfortable. Wear eye protection while working overhead, threading pipe, and driving nails.

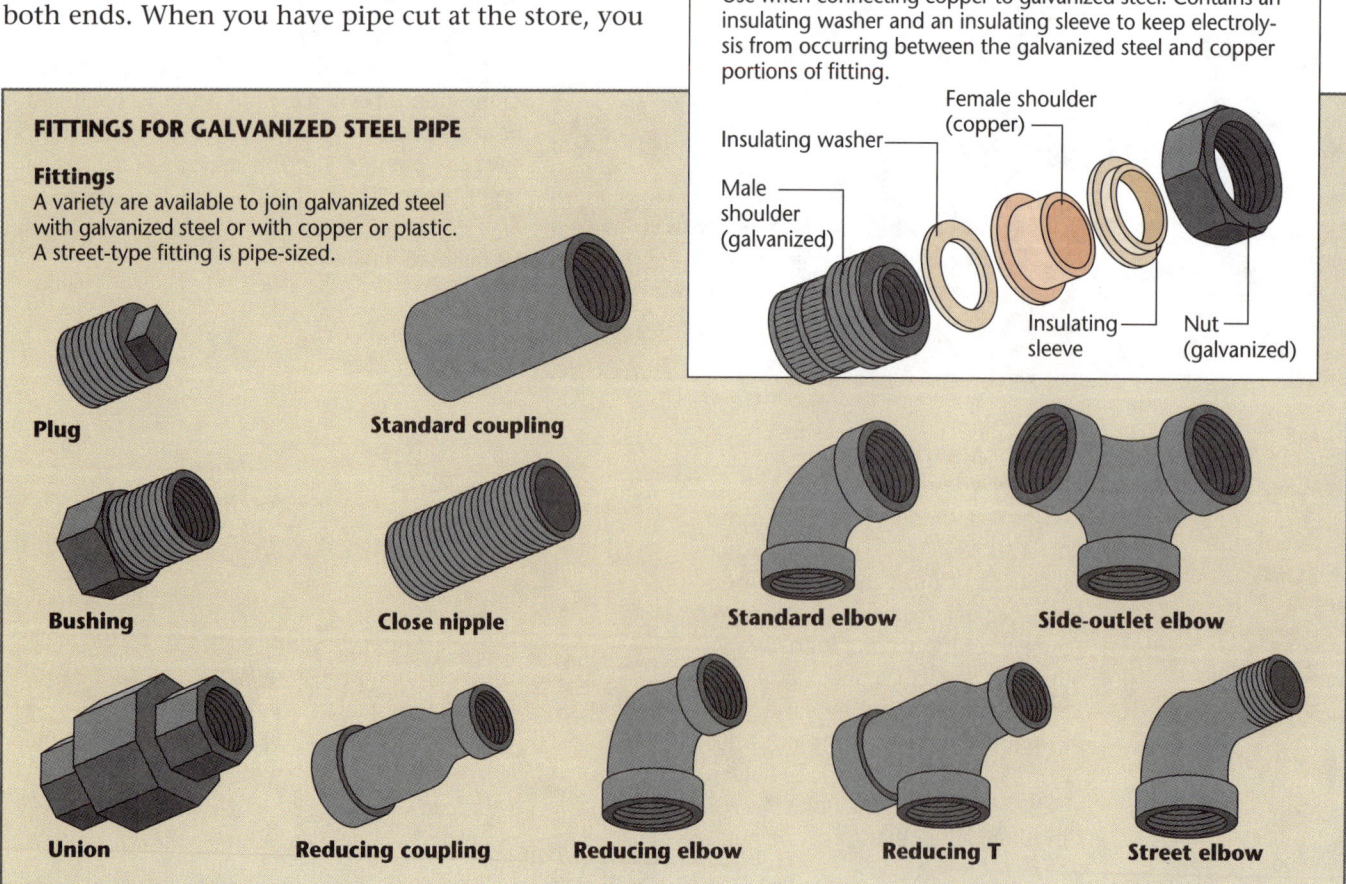

FITTINGS FOR GALVANIZED STEEL PIPE

Fittings
A variety are available to join galvanized steel with galvanized steel or with copper or plastic. A street-type fitting is pipe-sized.

Plug • Standard coupling • Bushing • Close nipple • Standard elbow • Side-outlet elbow • Union • Reducing coupling • Reducing elbow • Reducing T • Street elbow

Dielectric union
Use when connecting copper to galvanized steel. Contains an insulating washer and an insulating sleeve to keep electrolysis from occurring between the galvanized steel and copper portions of fitting.

Insulating washer • Female shoulder (copper) • Male shoulder (galvanized) • Insulating sleeve • Nut (galvanized)

USING PIPE WRENCHES

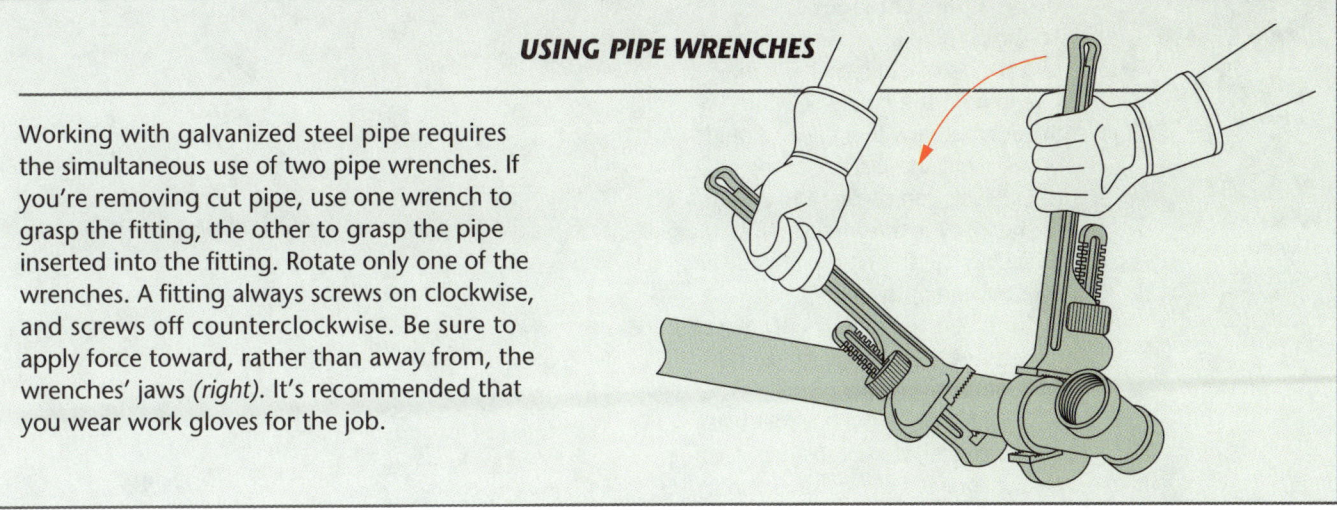

Working with galvanized steel pipe requires the simultaneous use of two pipe wrenches. If you're removing cut pipe, use one wrench to grasp the fitting, the other to grasp the pipe inserted into the fitting. Rotate only one of the wrenches. A fitting always screws on clockwise, and screws off counterclockwise. Be sure to apply force toward, rather than away from, the wrenches' jaws *(right)*. It's recommended that you wear work gloves for the job.

Removing galvanized steel pipe (supply only)

TOOLKIT
- Reciprocating saw or coarse-toothed hacksaw
- Pipe wrenches

Cutting and using two wrenches

If there's no union in the run, saw the pipe in two; unscrewing the pipe at one end before cutting would only tighten it at the other end. Hold the run steady with your hand or a wrench to stop excess motion that would strain joints; also keep cut ends from sagging. Use a reciprocating saw or a coarse-toothed (18 teeth per inch) hacksaw and put a bucket underneath to catch any spills.

Using 2 wrenches *(above)*, unscrew one section of the cut pipe, then the other. If the pipe sticks, apply liberal doses of penetrating oil to the joints; wait 5 minutes and unscrew.

PLAY IT SAFE

ALWAYS TURN OFF THE WATER
CAUTION: *Before beginning work on supply pipe, turn off the water supply at the house shutoff valve (page 10). Drain the pipes by opening a faucet at the low end. And be sure to wear safety goggles and work gloves when cutting or threading pipe.*

Measuring and cutting galvanized steel pipe

TOOLKIT
- Steel pipe cutter
- Pipe reamer

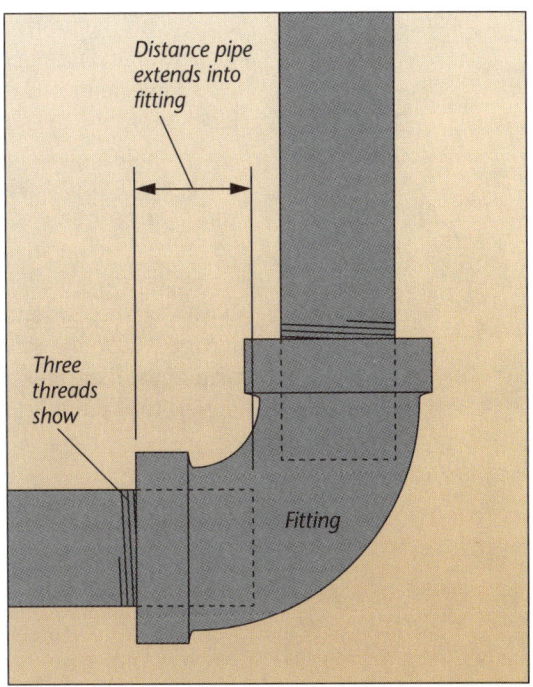

Measuring galvanized steel pipe
To determine exactly how much new galvanized steel pipe you will need, measure the distance between the new fittings, and then add the distances that the pipe is going to extend into the fittings; this is called makeup *(left)*. The distance allowed for each fitting should be less than the pipe's diameter.

Cutting galvanized steel pipe
It's important to cut galvanized pipe perfectly square so that threads can be accurately started in its new ends. Using a pipe cutter with a blade designed for steel pipe, follow the directions for cutting copper tube found on page 19. After you've finished cutting, use a retractable reamer in the cutter handle to remove burrs from the inside of the pipe (threading will remove any burrs from the outside surface).

GETTING STARTED IN PLUMBING

Threading galvanized steel pipe

TOOLKIT
- Pipe vise
- Pipe threader
- Wire brush

Using a pipe threader
To thread pipe at home you'll need two pieces of equipment: a pipe vise to hold the pipe steady, and a threader fitted with a die and guide of the same nominal size as the pipe (these tools can be rented). You fit the head of the threader (die) into the threading handle and slip it guide first over the end of the pipe.

To thread the pipe *(right)*, exert force toward the body of the pipe while rotating the handle clockwise. When the head of the threader bites into the metal, stop pushing and simply continue the clockwise rotation. Apply generous amounts of thread-cutting oil as you turn the threader. If the threader sticks, some metal chips are probably in the way; back the tool off slightly and blow the chips off.

Continue threading until the pipe extends about one thread beyond the end of the die. Remove the threader from the pipe by rotating counterclockwise and clean off the newly cut threads with a stiff wire brush.

Joining galvanized steel pipe

TOOLKIT
- Pipe wrenches

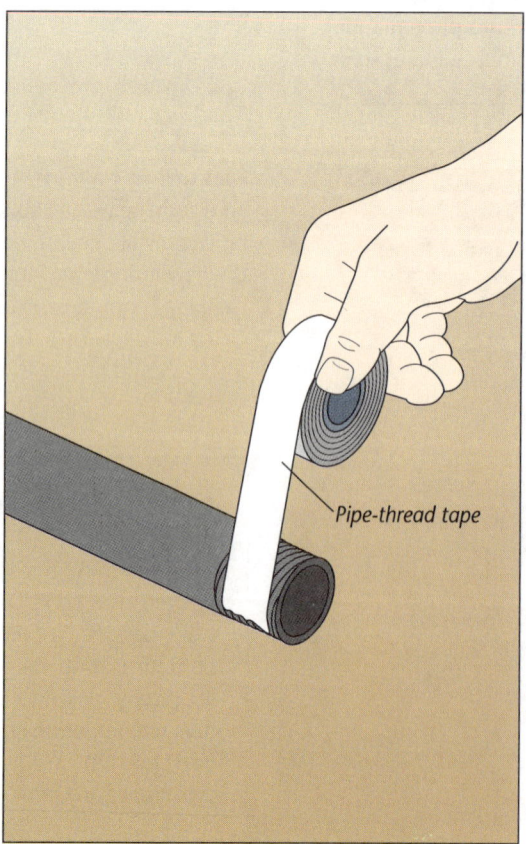

Using pipe tape or joint compound
The threads on galvanized steel pipe should be covered with pipe joint compound or pipe-thread tape to seal them against rust and to make assembly and disassembly easier. Apply pipe joint compound with the brush attached to the lid of the container, using just enough to fill the male threads. If you use pipe-thread tape instead, wrap $1^1/_2$ turns clockwise around the pipe threads, pulling the tape tight enough so that the threads show through. Do *not* attempt to coat the interior threads of the fitting with compound or tape. Screw the pipe and fitting together by hand as far as you can. After you've tightened the pipe and fitting by hand, finish with two pipe wrenches as explained previously on page 23. Do this slowly—done too fast, joining creates heat that causes the pipe to expand; later, the pipe may shrink and the joint loosen.

WORKING WITH CAST-IRON PIPE

If your home was built before 1970, there's a good chance that it has cast-iron piping. Cast-iron pipe is strong, resists corrosion, and is dense enough to be the quietest of all piping materials. Rubber joints reduce vibration and allow for expansion and contraction without additional fittings or insulation. While heavier than plastic, almost all cast-iron fittings used in home construction weigh less than four pounds. Since cast iron uses a gasketed joint, it's easily removed, reassembled, and used again.

There are two types of fittings for cast iron: hub-and-spigot and no-hub or hubless. Hub-and-spigot joints are usually found in older homes and might have been joined using molten lead and oakum. These materials are rarely used in residences now, as most codes no longer permit lead in DWV piping. The no-hub joint is most commonly used because it takes up little space—a 3-inch stack will fit into a 2x4 stud wall without extra preparation. It's also the easiest to install and can be used to modify hub-and-spigot systems.

A wide variety of fittings is available for use with both types of cast-iron pipe. Drainage fittings, unlike water-supply fittings, don't have interior shoulders, and each has a built-in fall or slope to allow for gravity flow.

Cast iron ranges in size from 1 1/2 inches to 4 inches and larger. It is usually bought in 10-foot lengths; short lengths are available for hub-and-spigot.

Cast iron needs only to be supported every 5 feet in horizontal runs, except where a full 10-foot length of pipe is used, then every 10 feet is adequate. Support should be located within 18 inches of the joint. Plumber's tape, hangers, and straps are commonly used.

CAUTION: Before beginning any work on cast-iron pipe, be sure to turn the water off at the house shutoff valve *(page 10)*.

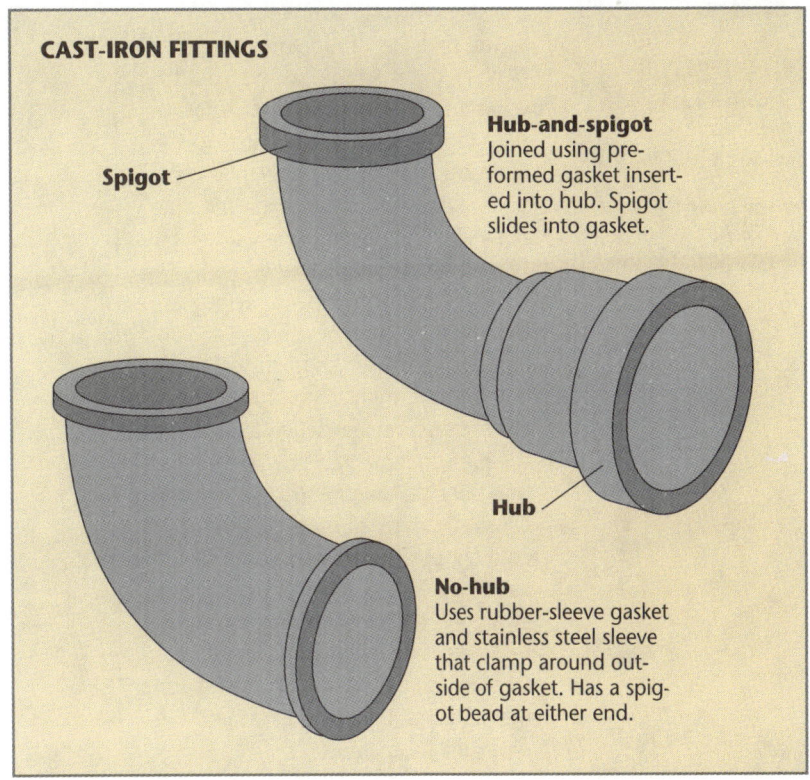

CAST-IRON FITTINGS

Hub-and-spigot Joined using preformed gasket inserted into hub. Spigot slides into gasket.

No-hub Uses rubber-sleeve gasket and stainless steel sleeve that clamp around outside of gasket. Has a spigot bead at either end.

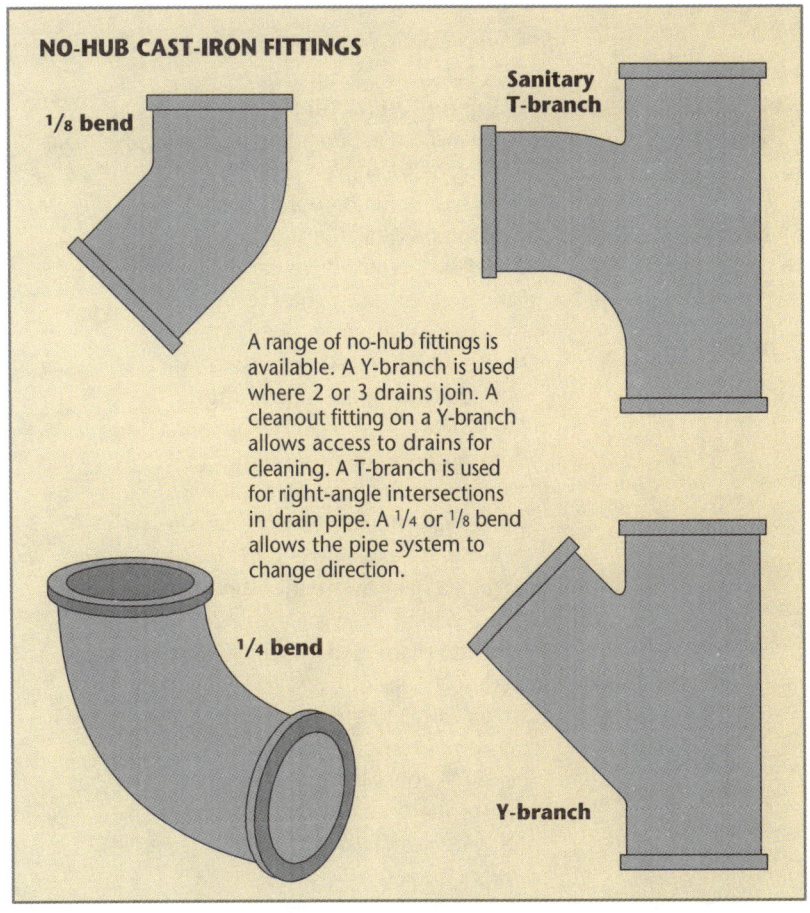

NO-HUB CAST-IRON FITTINGS

A range of no-hub fittings is available. A Y-branch is used where 2 or 3 drains join. A cleanout fitting on a Y-branch allows access to drains for cleaning. A T-branch is used for right-angle intersections in drain pipe. A 1/4 or 1/8 bend allows the pipe system to change direction.

GETTING STARTED IN PLUMBING

Removing cast-iron pipe

TOOLKIT
- Reciprocating saw
OR
- Cast-iron snap cutter
OR
- Hacksaw
- Cold chisel
- Ball-peen hammer

Cutting the pipe

When cutting cast-iron pipe, wear safety goggles and work gloves. Before removing pipe, securely support the section to be removed, as well as the sections of pipe on either side of it. Plumber's tape can be used to wrap and support the pipe. Pull the tape taut and nail it to nearby joists or studs. To determine how much new no-hub pipe you need, simply measure between the cut ends where a section of pipe has been removed.

To cut cast-iron pipe, a reciprocating saw is the tool of choice. Or, you can use a cast-iron snap cutter, available at equipment-rental stores. This cutter uses a ratchet action to increase the pressure equally on cutting wheels that encircle the pipe *(right)*, which causes the pipe to snap. A hacksaw, cold chisel, and ball-peen hammer can also be used: Chalk a cutting line all around the pipe, and then score it to a depth of 1/16" with the hacksaw. Deepen the cut with a ball-peen hammer and chisel, tapping all around the pipe until it breaks.

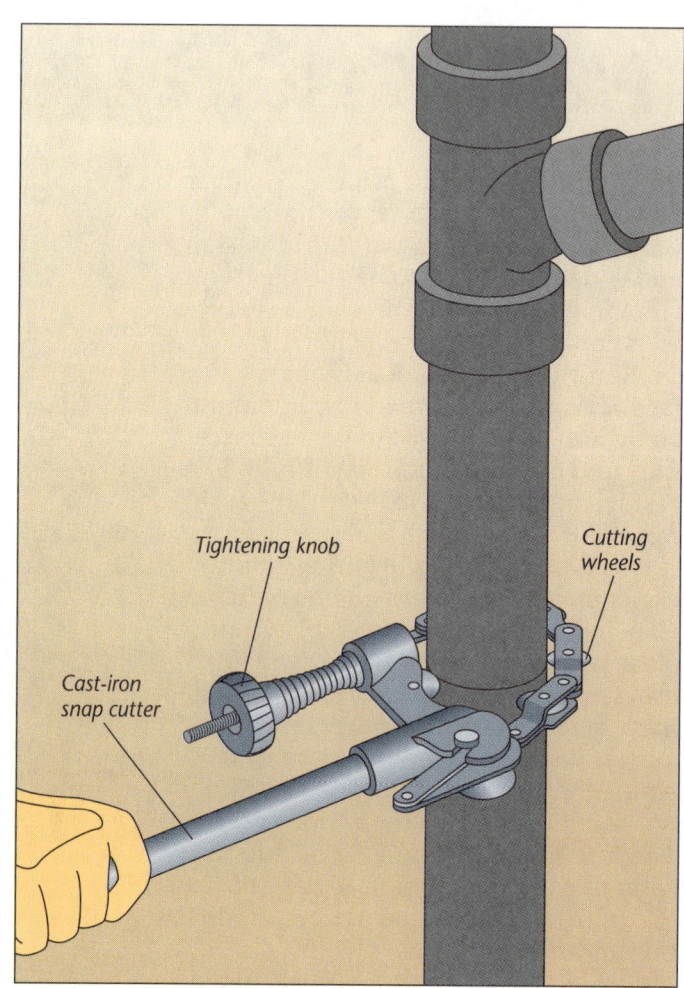

Joining cast-iron pipe

TOOLKIT
- Torque wrench

Using no-hub fittings

To connect a no-hub fitting or pipe to existing cast-iron pipe, a no-hub coupling is used. The coupling consists of a neoprene gasket, a stainless steel shield, and worm-drive band clamps for compressing the gasket around the pipe. To assemble this joint, slip the coupling and gasket assembly onto the end of the pipe or fitting *(right, above)*. The gasket sleeve has a built-in stop to help you center the assembly at the joint. The stainless steel shield should be moved above or below the opening and the gasket lip folded back before the fitting or pipe is added. Next roll the gasket lip back into place, slide the shield into place *(right, middle)*, position the band screws with a torque wrench: 60 inch-pounds of torque is required *(right, below)*. A special wrench preset for this torque should be available at an equipment-rental dealer.

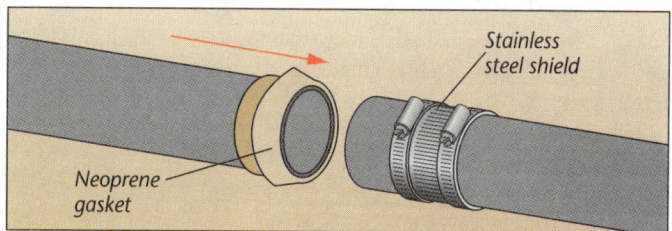

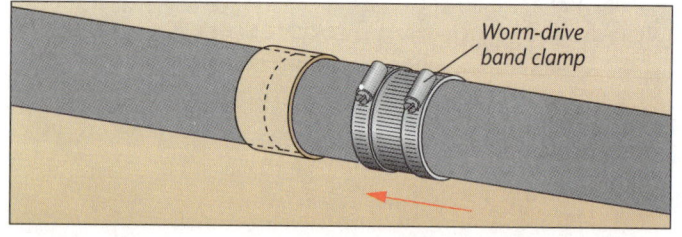

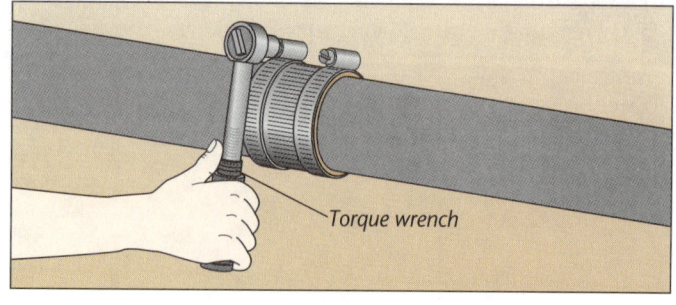

PLUMBING REPAIRS

This chapter will show you how to solve the most common plumbing problems, including leaking fixtures and defective pop-ups, frozen pipes, and low water pressure. A clogged drain, usually the most urgent situation, may also be the easiest to remedy. In most cases you don't need to reach for the phone to call a plumber; you can probably deal with the clog yourself. You'll be able to give this emergency immediate attention if you have drain-cleaning tools on hand—a plunger and a drain-and-trap auger.

In addition to being an annoyance, a leaking plumbing fixture can be a serious water-waster; a leaky faucet may use $1/3$ of a gallon per hour, and a running toilet five gallons per minute. It's definitely worth trying to repair a faucet, toilet, or valve before deciding to invest in an expensive replacement. Most plumbing fixtures can be repaired simply, using common tools. Parts are generally inexpensive, but in most cases you'll need to make sure they're exact replacements for your particular model. Sometimes, nothing more than a washer is required—it's a good idea to keep a selection on hand. Turn to page 8 for some of the tools and supplies you'll need for basic plumbing repairs. Remember to tape the jaws of your wrench or pliers when gripping visible parts of faucets and other fixtures to prevent scratches.

Many plumbing repairs necessitate turning off the water. You can make your life much easier by installing shutoff valves at all your fixtures, so you don't have to turn off the water supply to the whole house each time a problem arises. Detailed instructions for installing fixture shutoffs are given in the next chapter, on page 75. For more information on turning off the water supply, turn to page 10.

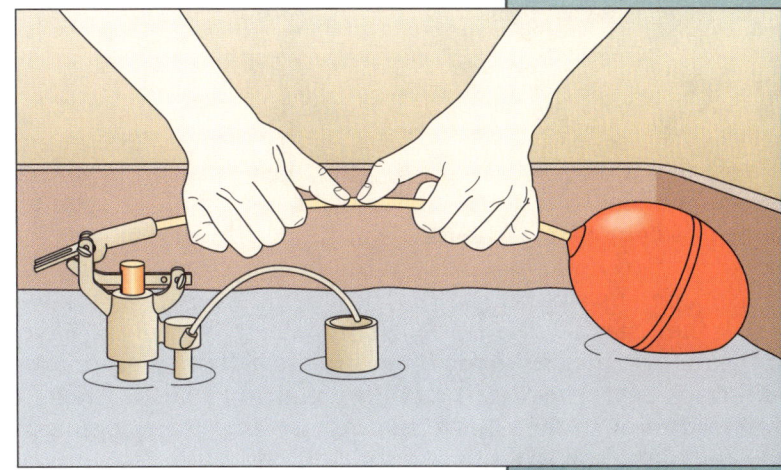

Toilet running? All that it may take to fix this is bending down the float arm, as shown here. In this chapter we'll also show you more complex toilet repairs such as replacing the inlet valve or flush valve.

CLEARING OUT CLOGS

In this section, you'll find out how to clear out clogged sinks, tubs and showers, and main drains. (Unclogging toilets is dealt with starting on page 53.)

The ideal is to prevent clogs entirely (see the information below). At least be alert to the warning sign of a sluggish drain—it's easier to open a drain that's slowing down than one that's stopped completely.

Usually, a clog will be close to the fixture. You can determine this by checking the other drains in your home. If only one fixture is blocked, you're probably dealing with a clog in the trap or drainpipe of that fixture. (In some cases a trap may need to be replaced—see page 61.)

If more than one drain won't clear, something is stuck farther along in a branch drain, the main drain, or the soil stack *(page 31)*, causing all the fixtures above the clog to stop up. If there's a blockage in the vent stack (the pipe that keeps air flowing out the roof), wastes drain slowly, and odors from the pipes become noticeable in the house. (See page 5 for a drawing of these systems.)

Chemical drain cleaners can be used preventively, on a regular basis, but should be avoided for clearing clogs; they leave you with caustic fluids to get rid of afterwards. Instead, rely on the tried-and-true weapons in the drain-cleaning arsenal: the plunger or the auger.

MAINTENANCE TIP

PREVENTING CLOGGED DRAINS

A kitchen sink usually clogs because of a buildup of grease and food particles that get caught in it. To keep the problem to a minimum, don't pour grease down the drain. Another villain is coffee grounds—throw them out, don't wash them down.

Hair and soap are usually at fault in bathroom drains. Clean out strainers and pop-ups regularly. Some strainers are held in place by screws. For removing sink and tub pop-ups, turn to page 46.

There are many chemical drain cleaners on the market; you can use one of them preventively (about once a month) before a clog forms. Read labels, and match cleaners with the material you're trying to dissolve. Alkalis cut grease; acids dissolve soap and hair. Before you pour the cleaner down the drain, block the overflow vent with rags. When using chemicals, follow these safety guidelines: Work in a well-ventilated room and wear rubber gloves and goggles. Don't mix chemicals; mixing an acid and an alkali cleaner can cause an explosion. Don't look down the drain after pouring in a chemical; the solution frequently boils up and gives off toxic fumes. Never plunge if you've recently used a chemical.

Consider using equal parts baking soda and vinegar instead of a commercial cleaner against soap and hair. First dump the baking soda down the drain, then the vinegar. Let the mixture fizz, then flush the drain with boiling water.

SINKS

A dose of scalding water is often effective against grease buildups. If not, it could be that some small object—a coin or small utensil—has slipped down the drain. To check, remove (and thoroughly clean) the sink pop-up stopper *(page 46)* or the strainer *(page 45)*. If these simple measures fail, try the sink plunger. If the plunger also fails, you'll need to use a drain-and-trap auger *(opposite)*, a flexible metal coil that you feed through the pipes until it reaches the clog; the end of the coil snags the clog and dislodges it or pulls it right out.

First, check for the clog near the drain by inserting the auger down through the drain. If that doesn't clear the clog, put the auger in through the trap cleanout, if there is one. And if that doesn't work, remove the trap entirely so the auger can reach through the drainpipe to clear a clog that's farther from the drain. Once you remove the trap, though, you may find you're able to solve the problem simply by cleaning out the trap with a flexible brush, such as a bottle brush, and soapy water.

If your sink has a garbage disposer and the disposer drainpipe clogs, disassemble the trap and thread an auger into the drainpipe. If both basins of a double sink with a garbage disposer clog, snake down from the one without a disposer. If only the basin with the disposer is clogged, you'll have to remove the trap to dislodge the blockage.

If an auger through the sink drainpipe doesn't succeed, the clog is probably too deep in the pipes to reach through the drainpipe. This means you're dealing with a main drain clog—see directions on page 31.

28 PLUMBING REPAIRS

Using a plunger to clear a clog

TOOLKIT
- Sink plunger

Efficient plunging
When using a plunger to clear a drain, don't make the typical mistake of pumping up and down only 2 or 3 times, expecting the water to whoosh down the drain. Choose a plunger with a suction cup large enough to cover the drain opening completely. Fill the clogged fixture with enough water to cover several inches of the plunger cup. Then, use a wet cloth to block off all other outlets (the overflow vent, the second drain in a double sink, adjacent fixtures) between the drain and the clog. Coat the rim of the plunger cup with petroleum jelly to ensure a tight seal, and insert the plunger into the water at an angle so that little air remains trapped under it. Use 15 to 20 forceful strokes, holding the plunger upright *(right)*; the last stroke should be a vigorous upstroke, snapping the plunger off the drain and hopefully drawing the clog with it. Repeat the process 2 or 3 times before giving up.

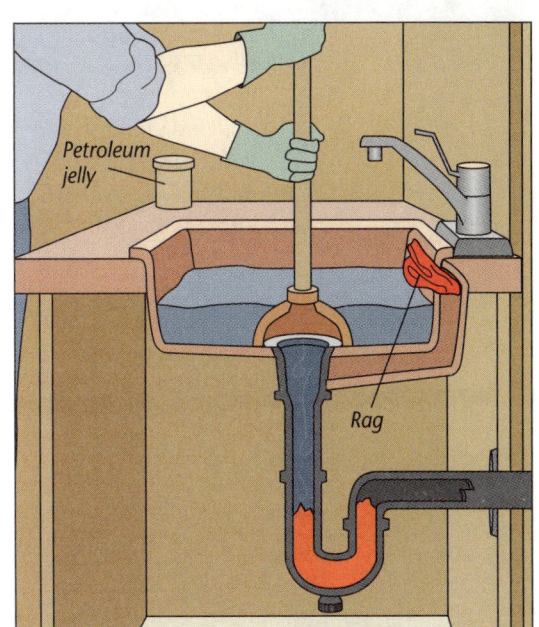

HOW TO USE A DRAIN-AND-TRAP AUGER

Feed the auger into the drain, trap, or pipe until it stops. If there is a movable hand grip, position it about six inches above the opening and tighten the thumbscrew. Rotate the handle clockwise to break the blockage. (Never rotate counterclockwise, as it can damage the cable.) As the cable works its way into the pipe, loosen the thumbscrew, slide the hand grip back, push more cable into the pipe, tighten again, and repeat. If there is no hand grip, push and twist the cable until it hits the clog. The first time the auger stops, it probably has hit a turn in the piping rather than the clog. Guiding the auger past a sharp turn takes patience and effort; keep pushing it forward, turning it clockwise as you do. Once the head of the auger hooks the blockage, pull the auger back a short distance to free some material from the clog, then push the rest on through.

After breaking up the clog, pull the auger out slowly and have a pail ready to catch any gunk that is brought out. Flush the drain with hot water. Dry the auger and coat it with a lubricant before putting it away.

Using an auger to clear a clog

TOOLKIT
- Drain-and-trap auger
- Adjustable wrench

Going through the drain
Remove the pop-up stopper *(page 46)*, and the sink strainer *(page 45)*. Insert the auger in the drain opening and twist it down through the trap until you reach the clog *(right)*.

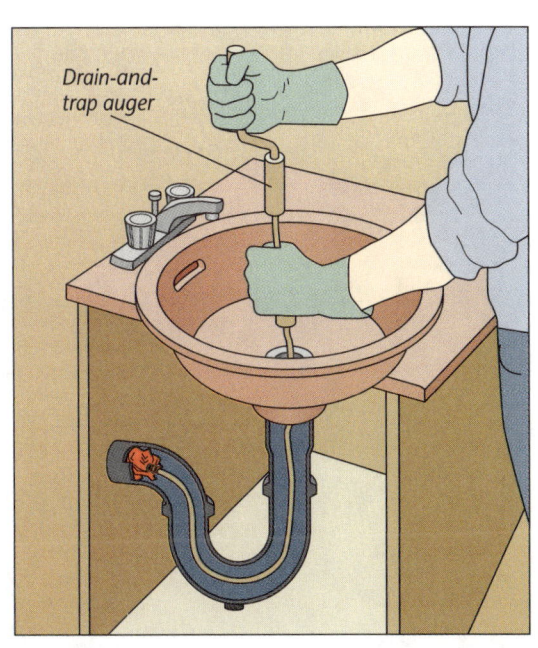

Going through the cleanout
If the trap has a cleanout, place a pail under it, and remove the cleanout plug from the bottom of the trap. Insert the auger (or a bent wire coathanger), direct it toward the drain, or angle it to reach a deeper blockage *(left)*.

PLUMBING REPAIRS

Going through the drainpipe
Remove the trap as outlined on page 61. Pull the trap downward and spill its contents into a pail. Insert the auger into the drainpipe at the wall. Feed it as far as it will go, turning clockwise until it hits the clog *(right)*. Clean out the trap before reinstalling it.

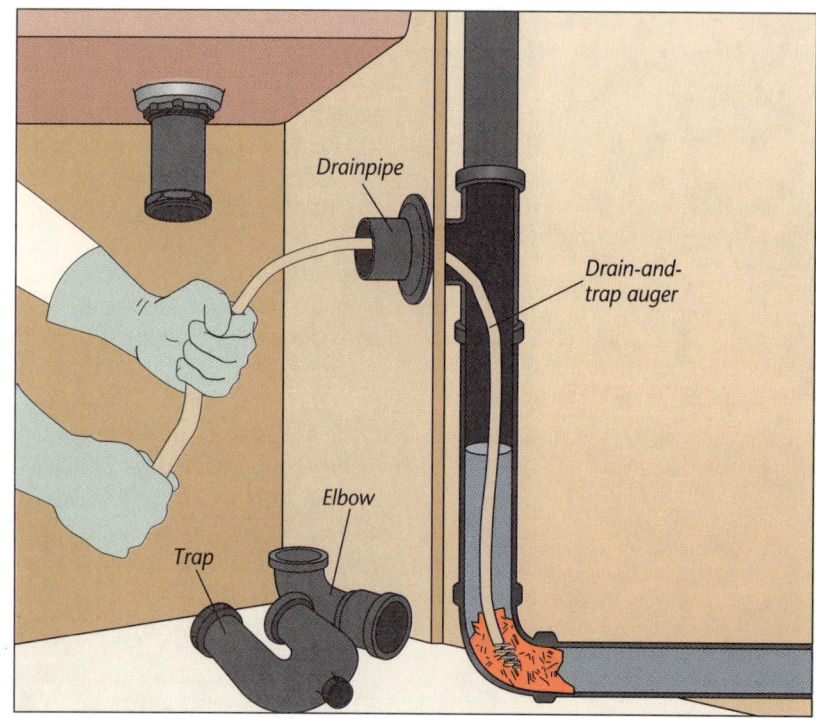

TUBS AND SHOWERS

To help prevent clogs in tubs and showers, you can install a hair trap in your tub. One type of hair trap simply sits in the drain; another requires replacing the pop-up.

When a tub or shower drain does clog up, first see whether other fixtures are affected. If they are, work on the main drain *(opposite)*. If only the tub or shower is plugged, work on it. First, try plunging *(page 29)*, then remove the strainer or pop-up and clean it *(page 47)*. If this doesn't work, use an auger *(page 29)* or a balloon bag *(page 32)*; insert the bag past any other openings in the drain.

Unclogging a tub drain

TOOLKIT
- Drain-and-trap auger

For P-trap:
- Screwdriver

For drum trap:
- Adjustable wrench

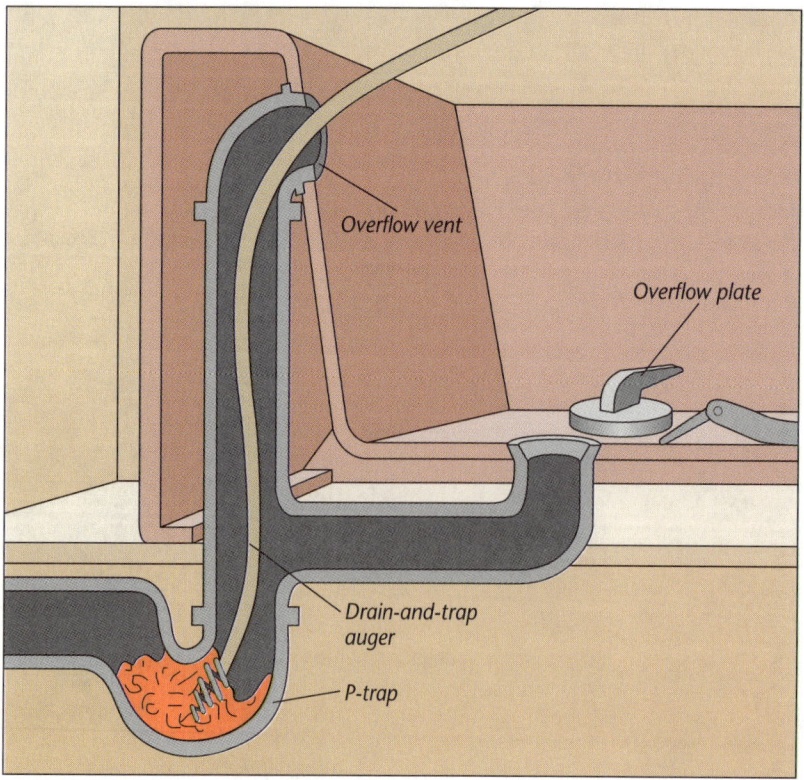

Working on a drain with a P-trap
Remove the overflow plate and pull the pop-up or plunger assembly *(page 47)* out through the opening. Feed the auger down through the overflow pipe and into the P-trap *(left)*. Use the twisting, probing method described on page 29. This should clear the drain. If not, and if possible, remove the trap or its cleanout plug from below or through an access panel *(page 61)*; have a pail ready to catch water. Insert the auger toward the main drain.

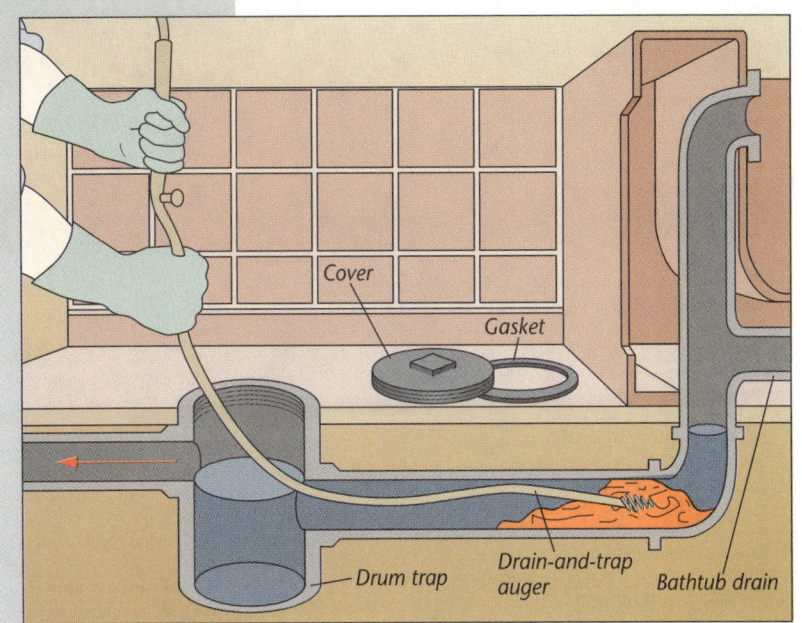

Working on a drain with a drum trap
Instead of a P-trap, bathtubs in older houses may have a drum trap located alongside the tub. If there's a clog, bail all water from the tub. Slowly unscrew the drum trap cover with an adjustable wrench. Watch for any water welling up around the threads; have rags ready. Remove the cover and rubber gasket on the trap and clean any debris from the trap. If you find no obstruction there, work the auger through the lower pipe toward the tub *(left)*. Still no clog? In that case, direct the auger in the opposite direction toward the main drain.

Unclogging a shower drain

TOOLKIT
- Drain-and-trap auger or garden hose
- Screwdriver (optional)

Using an auger or hose
Unscrew the strainer if your auger can't be threaded through it. Probe the auger down the drain and through the trap until it hits the clog *(right)*.

You can also use a garden hose instead of an auger. Use a threaded adapter to attach the hose to a faucet or run it to an outside hose bibb. Push the hose deep into the drain trap and pack wet rags tightly into the opening around it *(inset)*. Hold the hose and rags in the drain, and turn the water to the hose alternately on full force and abruptly off. Never leave a hose in a drain; a sudden drop in water pressure could siphon raw sewage back into the fresh-water supply. Be sure to clean and disinfect the hose before using it for other purposes.

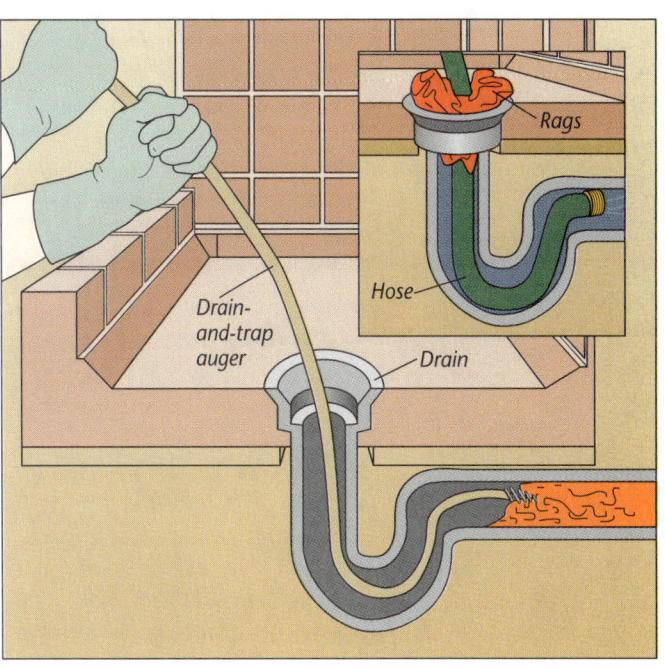

MAIN DRAINS

If a clog is too deep in the pipes to get at from a fixture, you can clean out the soil stack from below by working on a branch cleanout, the main cleanout, or the house trap (see the diagram on page 5). Cleaning the soil stack from below means working with raw sewage; have rubber gloves, pails, mops, and rags on hand.

First try a hand-operated drain-and-trap auger; turn to page 29 for instructions. If you're trying to clear the clog by working through a cleanout, you can use a balloon bag attached to a hose nozzle *(page 32)* instead. If neither of these methods does the job, you can choose to rent a power auger. Know your drain's diameter when you go to rent an auger. Use caution and work with a friend; be sure to ask for explicit safety instructions, and always plug into a GFCI-protected outlet.

If you don't want to try using a power auger, it's time to call a plumber or professional drain-cleaning firm. They may decide to try cleaning the soil stack through the vent stack from the roof; because of the danger, this should only be done by a professional.

PLUMBING REPAIRS

Working on the main or branch cleanout

TOOLKIT
- Pipe wrench
- Drain-and-trap auger
- Balloon bag and hose (optional)

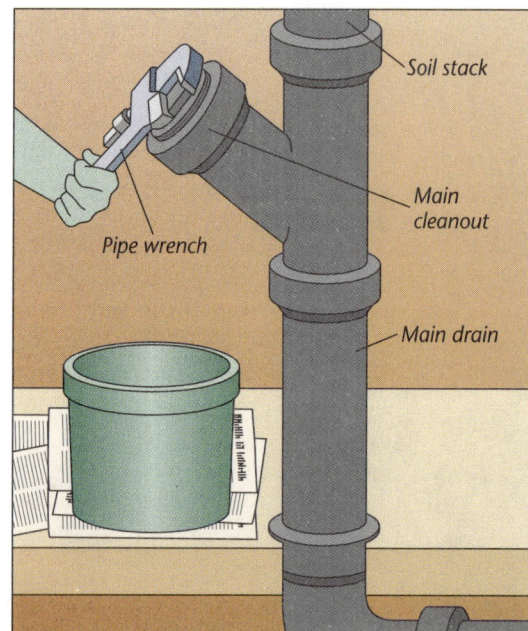
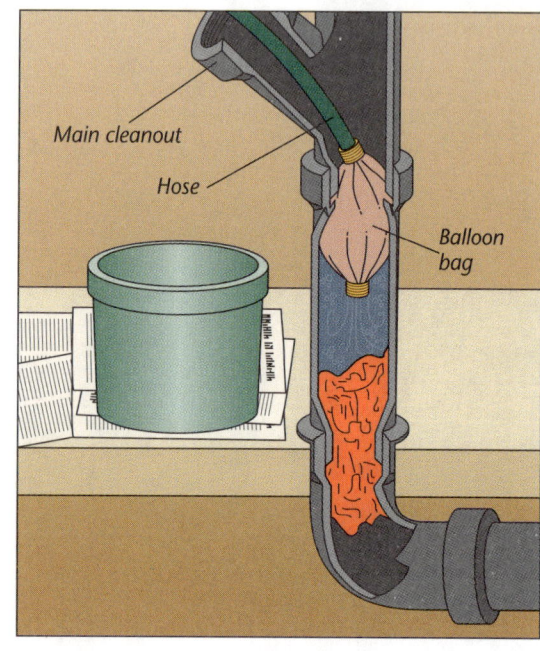

Opening and clearing the pipe

The main cleanout is usually a Y-shaped fitting located near the bottom of the soil stack where the main drain leaves the house. This is generally in the basement or crawl space, or on an outside wall near a toilet. In most newer buildings, there are also several branch cleanouts; try reaching the clog from one of these first.

Put on rubber gloves and set up a pail and newspapers to catch the waste water. Then, slowly remove the plug with a pipe wrench *(above, left)*. Try using an auger *(page 29)*. If this doesn't work, try a balloon bag and hose *(above, right)*; buy a bag that matches the diameter of your drain. Never leave the hose in the drain, and be sure to clean and disinfect it afterwards.

Once the drain is clear, flush it with water. Coat the plug with pipe-joint compound and recap the cleanout. If you are unable to clear the drain, move downstream to the house trap.

Unclogging the house trap

TOOLKIT
- Pipe wrench
- Drain-and-trap auger
- Wire brush

Probing the trap

The house trap is located in the basement, crawl space, or yard near where the main drain leaves the house. Two adjacent cleanout plugs are extended up to floor or ground level.

Before opening the house trap, put on rubber gloves and spread newspapers or rags around the cleanout. With a pipe wrench, slowly loosen the plug nearest the outside sewer line. Probe the trap and its connecting pipes with an auger *(right)*. (See page 29 for instructions on using an auger.) Be prepared to withdraw the snake and cap the trap quickly when water starts to flow. When the flow subsides, open both ends of the trap and clean it out with an old wire brush. Recap and flush the pipes with water from an upstream cleanout.

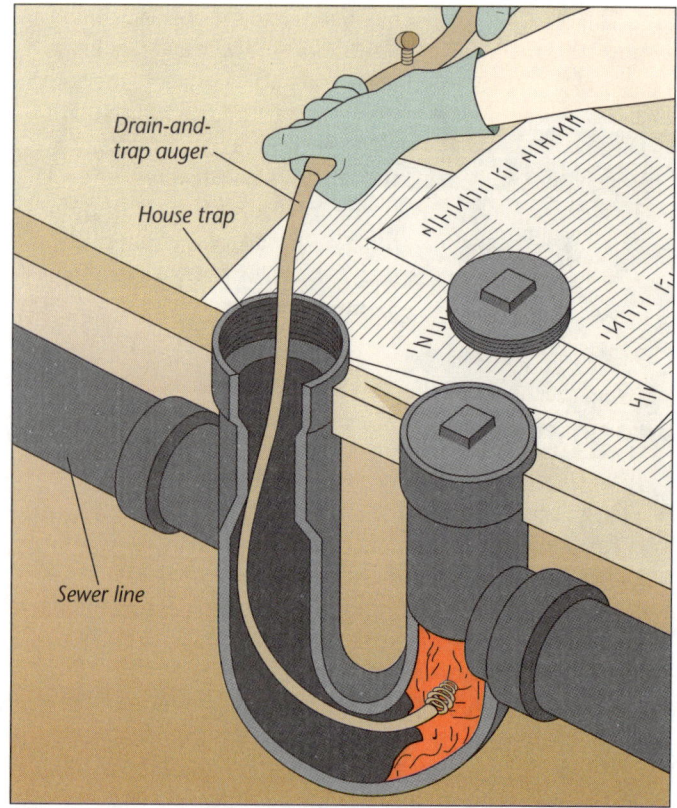

FIXING FAUCETS

This section will show you how to repair each type of faucet in turn. The faucets shown in the chapter are sink faucets; tub faucets are most often of the reverse-compression type, which can be repaired in the same way as a compression type or ball type. If a faucet isn't repairable, the ultimate solution is to replace it; we tell you how in the chapter that begins on page 66. In addition to repairing faucets, this section includes instructions on repairing sink sprayers *(page 43)* and shower heads *(page 44)*.

The first step in servicing a faucet is cleaning the aerator. Almost every faucet has, at the tip of its spout, an aerator that mixes air and water for a smooth flow. You should clean aerators periodically to remove mineral and debris buildup, as explained below.

Before beginning work on a faucet, turn off the water supply at the fixture shutoff valves or, if there aren't any, at the house shutoff valve *(page 10)*. Then open the faucet to drain the pipes.

If the faucet itself needs repair, you must first identify the type you're dealing with. There are two basic types of faucets. The first is called a compression, stem, or washer faucet, and usually has two handles and one spout. This type of faucet closes by a screw compressing a washer against a valve seat. You can tell you have a compression faucet if the handle has a "spongy" feel—allowing you to continue tightening—when you close it. Compression faucets are available with a number of types of handles.

The other type of faucet is called a washerless or non-compression faucet. Here, the mix of hot and cold water is often controlled by one handle or knob by aligning interior openings with the water inlets (washerless faucets can also have two handles). Types of washerless faucets include disc, rotating-ball, and cartridge. As opposed to compression faucets, washerless faucets have a hard, definite close that doesn't allow further tightening. It is sometimes possible to tell which type of washerless faucet you have by the handle, but to be sure, you'll have to take the faucet apart and compare with the illustrations given in this section.

Depending on the type of faucet, typical problems include dripping from the handles themselves, dripping from the spout, or dripping from under the base of the spout. Even if it's the spout that's dripping, it's the handle or handles that need repair. Before starting any faucet repair, plug the sink drain so that small parts can't fall down it, and line the sink with a towel to prevent damage from parts or tools accidentally dropped. Line up disassembled pieces so that you'll be able to put them back together in the right order.

Servicing a faucet aerator

TOOLKIT
- Toothbrush
- Pin or toothpick

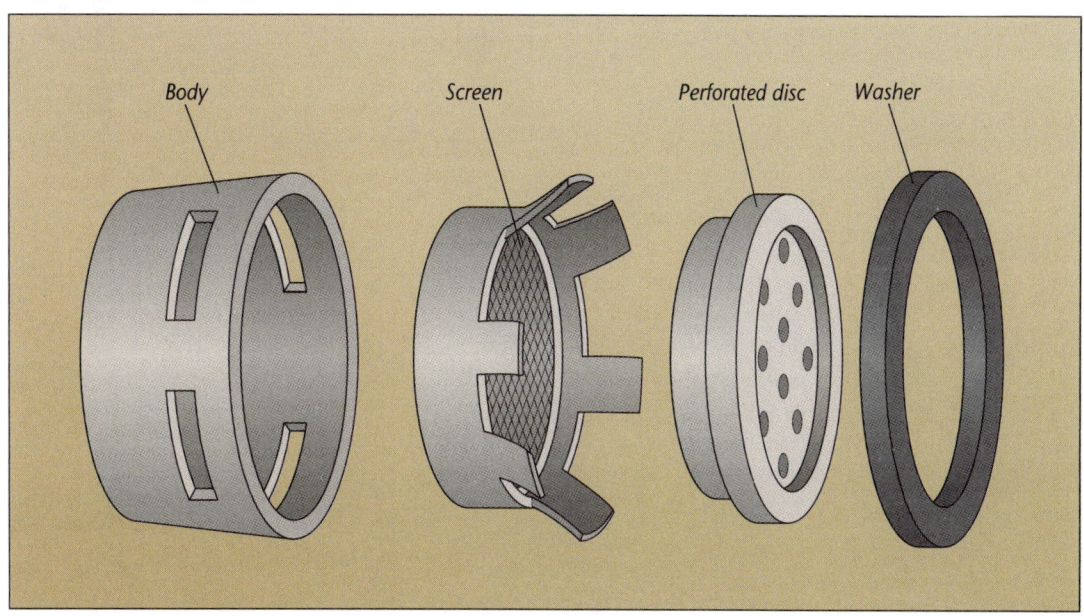

Cleaning the parts
Unscrew the aerator from the end of the spout or spray nozzle. Disassemble and set the parts aside—in order—for easy reassembly. Clean the screens and disc with an old toothbrush and soapy water; use a pin or toothpick to open any clogged holes in the disc. Hard-water scale can be removed by soaking the parts in vinegar or lime-dissolver. Flush all parts with water before putting them back together. If the parts are worn, it is simplest to replace the entire aerator.

PLUMBING REPAIRS 33

COMPRESSION FAUCETS

Ideally, when a compression faucet is turned off, the stem is screwed all the way down and the washer fits snugly onto the valve seat, stopping the flow of water. The packing nut compresses the packing—material that prevents water from being pushed up through the handle. If the faucet leaks from around the handle stem, you will need to either tighten the packing nut or replace the packing. If the spout is dripping, the problem is also with one of the handles; you'll need to replace the washer or seat or smooth the seat using a seat dresser. (If your faucet has a hat-shaped diaphragm instead of a washer, replace the diaphragm; however, these parts may be difficult to find.) To determine which handle needs work, turn off the shutoff valves under the fixture one at a time. The leak will stop when one or the other of them is turned off, and you will have narrowed down the problem.

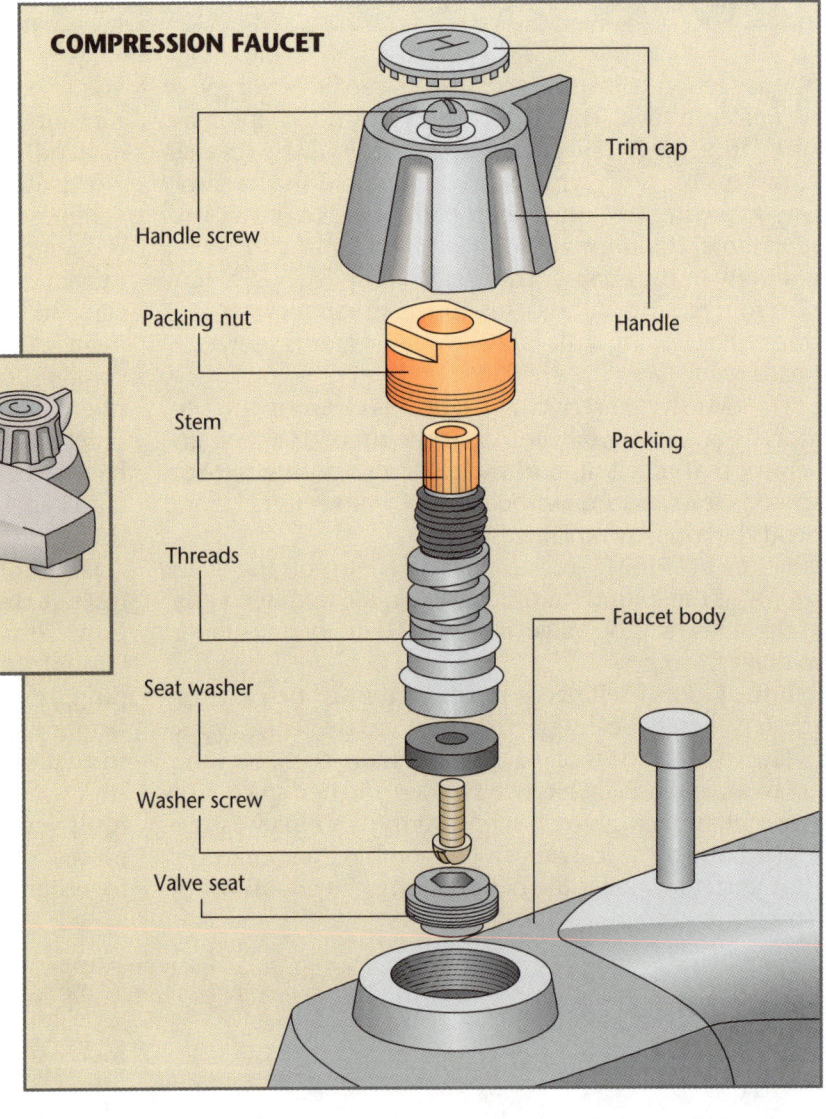

Repairing a leaking handle

TOOLKIT
- Blunt knife (optional)
- Screwdriver
- Adjustable wrench

1. Disassembling the faucet

To take the faucet apart, remove the trim cap on top of the handle, using a blunt knife or screwdriver. Undo the handle screw and pull or pry the handle straight up off its stem. Now tighten the packing nut one-quarter turn with an adjustable wrench. Reassemble the handle and turn the water back on.

If the leak persists, you'll need to replace the packing: Start by removing the packing nut with an adjustable wrench *(right)*.

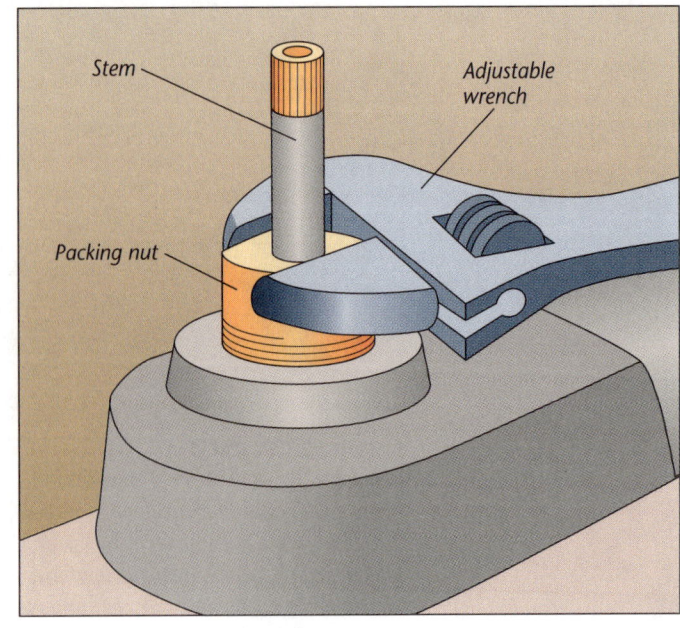

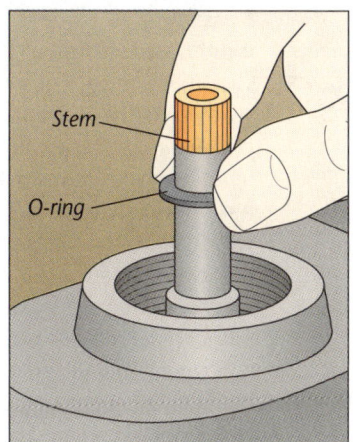
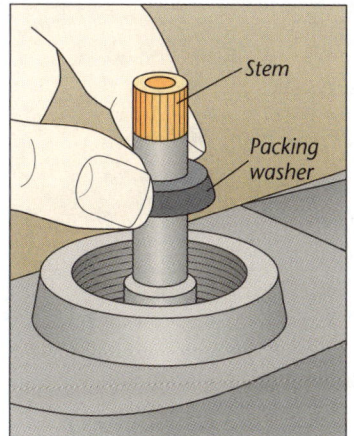
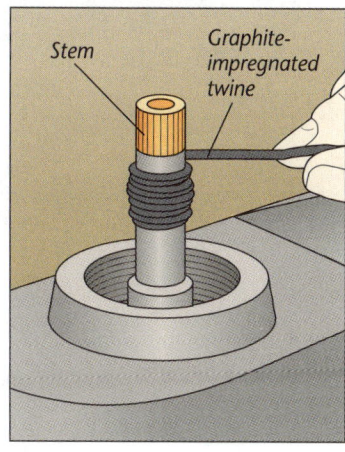

2 **Replacing the packing**

The packing on the faucet stem is either a rubber O-ring, a packing washer, or graphite twine. If it's an O-ring, pinch the old one off with your fingers *(above, left)*. Roll on an exact duplicate; lubricate it first with silicone grease. If it's a packing washer, remove the old one *(above, middle)* and push an exact replacement onto the stem. If it's graphite-impregnated twine, scrape away all of the old material and wrap new twine *(above, right)*, pipe-thread tape, or TFE packing clockwise—five or six times—around the faucet stem. Before replacing the packing nut, lubricate the threads with silicone grease. Tighten the packing nut and replace the handle.

Repairing a dripping spout

TOOLKIT
To replace seat washer:
- Screwdriver
- Adjustable wrench
- Drill and vise (optional)
- File (optional)

To service the valve seat:
- Valve-seat wrench or hex wrench
OR
- Valve-seat dresser

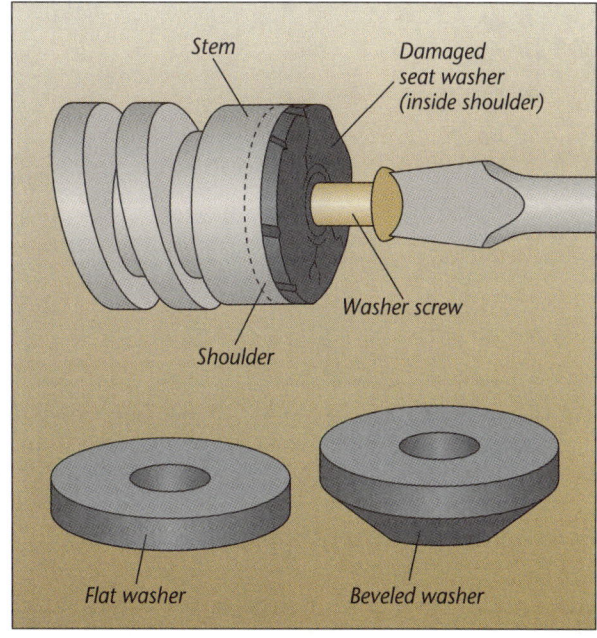
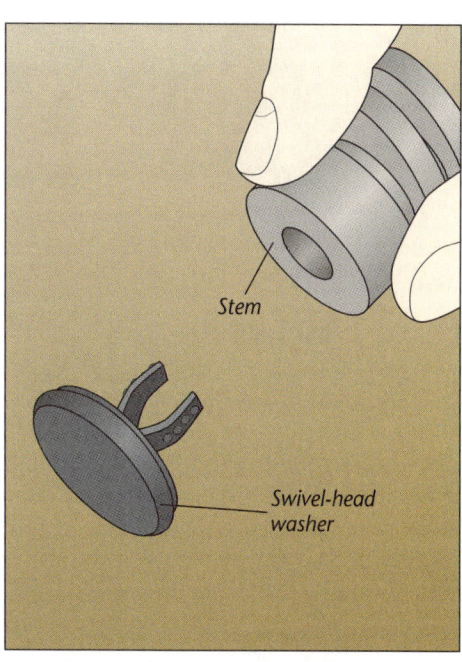

Replacing the seat washer

A spout leak is usually the result of a deteriorated washer or worn valve seat in the faucet. Remove the handle. Use the handle to turn the stem beyond its fully open position; then, remove the stem. At the bottom of the stem is a brass washer screw or seat screw that goes through the center of a rubberlike seat washer. If the washer is cracked, grooved, or marred, carefully remove the screw *(above, left)* and replace the washer with a new, identical one. Seat washers may be flat or beveled, as shown above. If the washer is beveled, be sure the beveled edge faces the screw head when you install it on the stem. During replacement, the shank of the washer screw may break off. If it does, you can replace it with a new swivel-head washer or replace the entire faucet stem. A swivel-head washer has two prongs that snap into the bottom of the stem to compress against the valve seat *(above, right)*. To install one, drill a hole in the end of the stem to receive the prongs. If the stem has a shoulder surrounding the washer, the shoulder will have to be filed away before a swivel-head washer can be installed.

PLUMBING REPAIRS 35

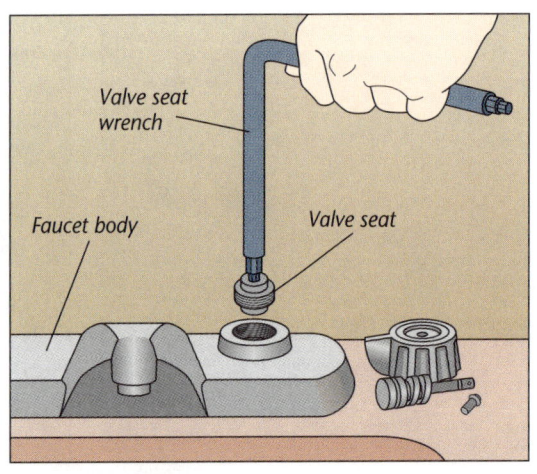

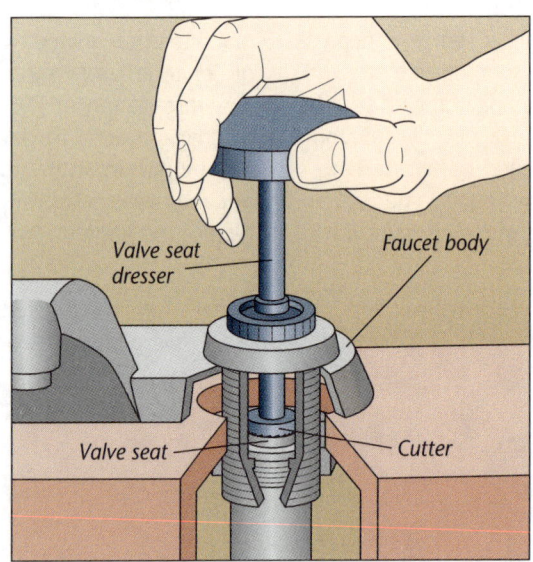

Replacing or smoothing the valve seat

If the washer isn't the problem, a damaged valve seat could be causing the leak by preventing the seat washer from fitting properly. On most compression faucets the valve seat is replaceable. You'll need a valve seat wrench—or the correct size of hex wrench—to make the exchange. Insert the wrench into the faucet body and turn it counterclockwise to remove the seat *(left, above)*. Buy an exact duplicate of the seat; before installing it, lubricate its threads with silicone grease.

If the faulty valve seat is built into the faucet, use a simple, inexpensive tool called a valve seat dresser to grind down any burrs on the seat, making it level and smooth. Seat-dressing tools usually come with a variety of cutter sizes; install the largest cutter that will fit into the faucet body. Insert the valve seat dresser until the cutter sits on the valve seat; then turn the tool handle clockwise until the seat is smooth *(left, below)*. Remove the metal filings with a damp cloth.

After the new seat washer or valve seat is in place, be sure to lubricate the threads of the stem with silicone grease before putting it back. Turn on the water, open the valve fully, and let it run for 10 seconds to flush out any cuttings left from dressing the seat.

DISC FAUCETS

In these washerless faucets, openings in two discs inside a sealed cartridge or stem unit assembly line up with inlet holes to allow the flow of water. In a single-handle type, the handle also controls the mixing of hot and cold. The cartridge in the single-handle model rarely wears out, but the stem unit assembly on a two-handle model may need to be replaced. More often, an inlet seal is the weak point. The single-handle type's three rubber seals provide for hot, cold, and mixed water. The two-handle type has a rubber or plastic seal and a small spring on each side. Disc faucets are available with a variety of handles, as shown below.

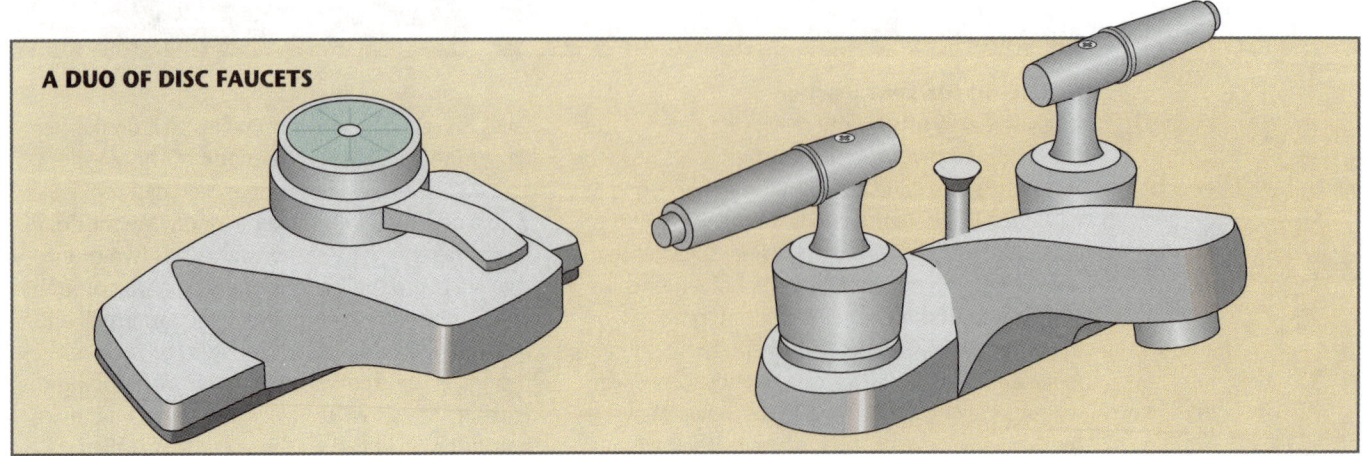

A DUO OF DISC FAUCETS

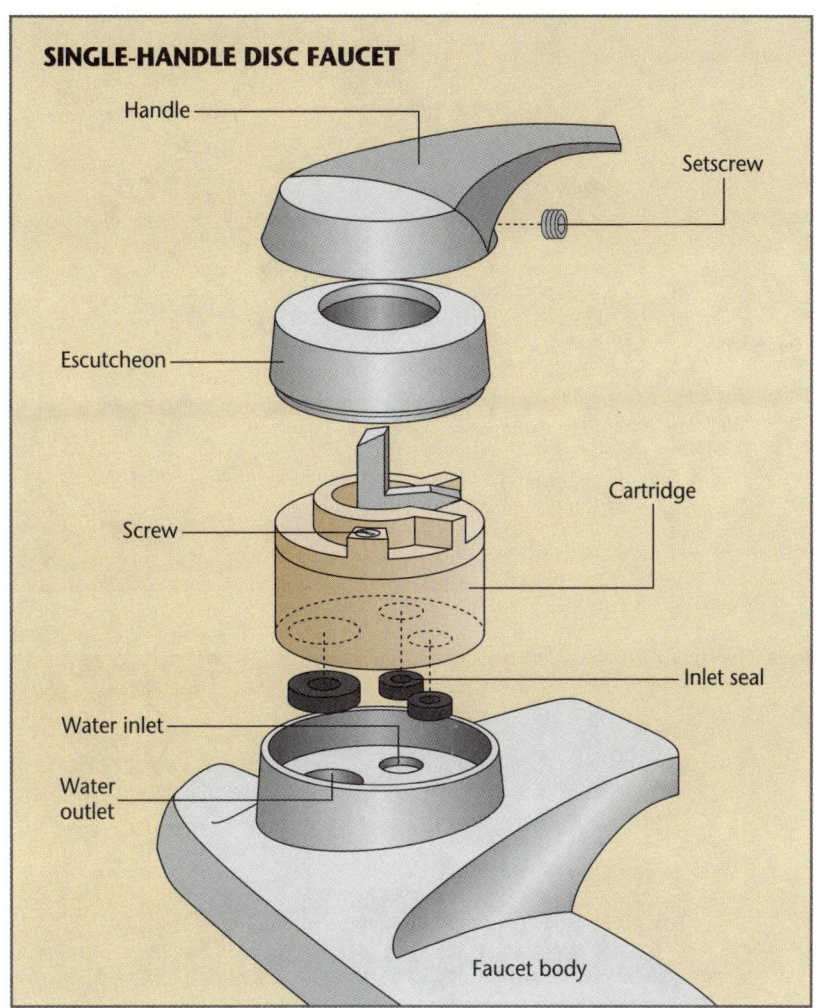

SINGLE-HANDLE DISC FAUCET

- Handle
- Setscrew
- Escutcheon
- Screw
- Cartridge
- Inlet seal
- Water inlet
- Water outlet
- Faucet body

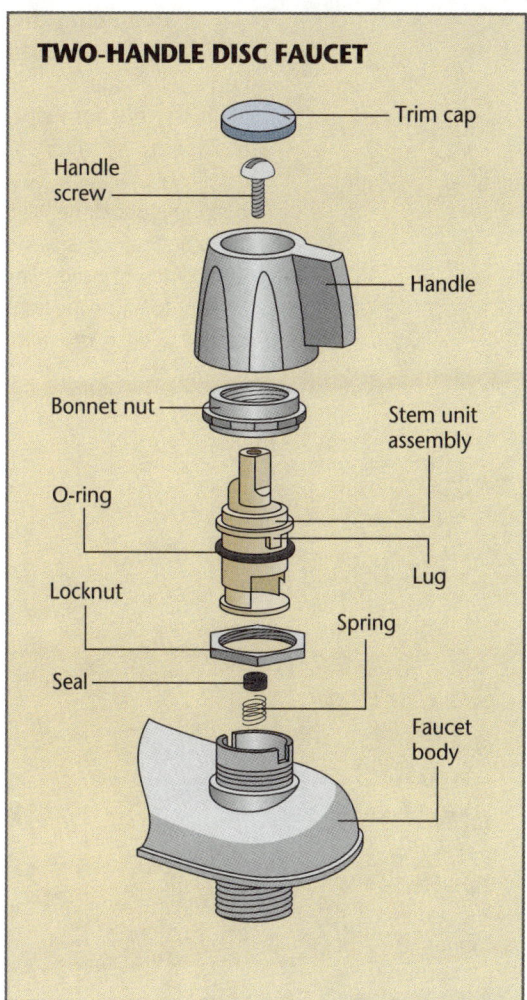

TWO-HANDLE DISC FAUCET

- Trim cap
- Handle screw
- Handle
- Bonnet nut
- Stem unit assembly
- O-ring
- Lug
- Locknut
- Spring
- Seal
- Faucet body

Repairing a single-handle disc faucet

TOOLKIT
- Screwdriver
- Long-nose pliers

1 ▶ Removing the cartridge

To repair a dripping spout or a leak at the base of a disc faucet, remove the setscrew under the faucet handle and lift off the handle and decorative escutcheon. Then remove the cartridge by loosening the two screws that hold the cartridge to the faucet body *(right)*.

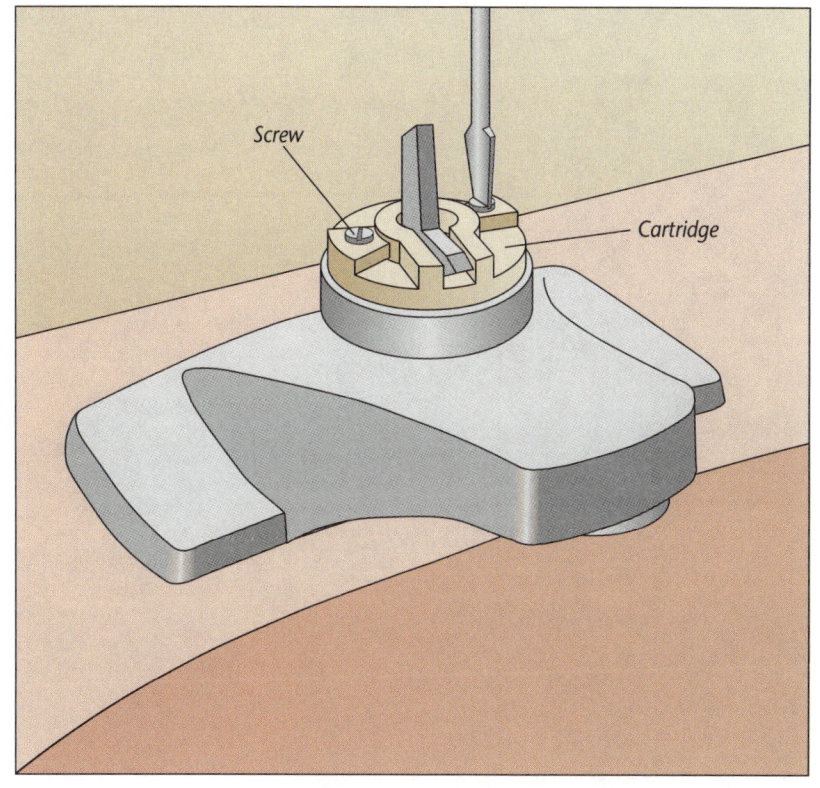

- Screw
- Cartridge

PLUMBING REPAIRS 37

2 Replacing the seals
Under the cartridge, you'll find a set of inlet seals *(right)*. Take each one out for inspection and replace any worn ones with exact duplicates. Also check for sediment buildup around the inlet holes; scrape away any deposits to clear the restriction. When reassembling the faucet, be sure to align the inlet holes of the cartridge with those in the base of the faucet.

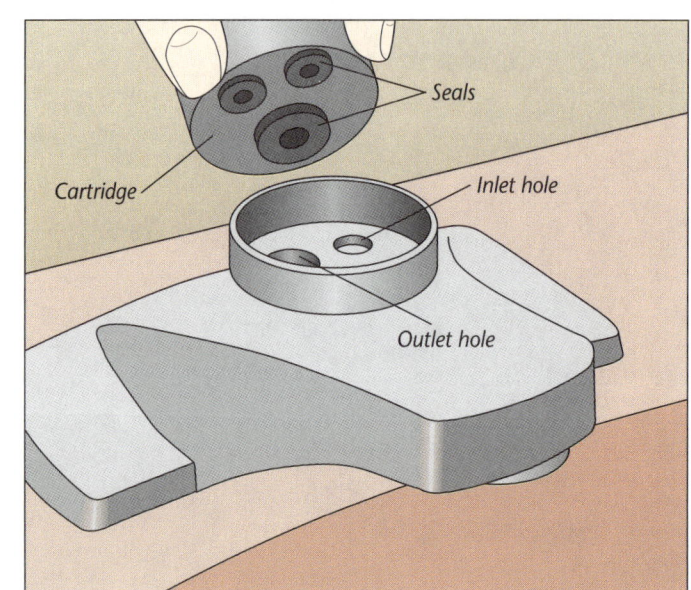

Repairing a two-handle disc faucet

TOOLKIT
- Blunt knife (optional)
- Screwdriver
- Adjustable wrench
- Rib-joint pliers
- Long-nose pliers

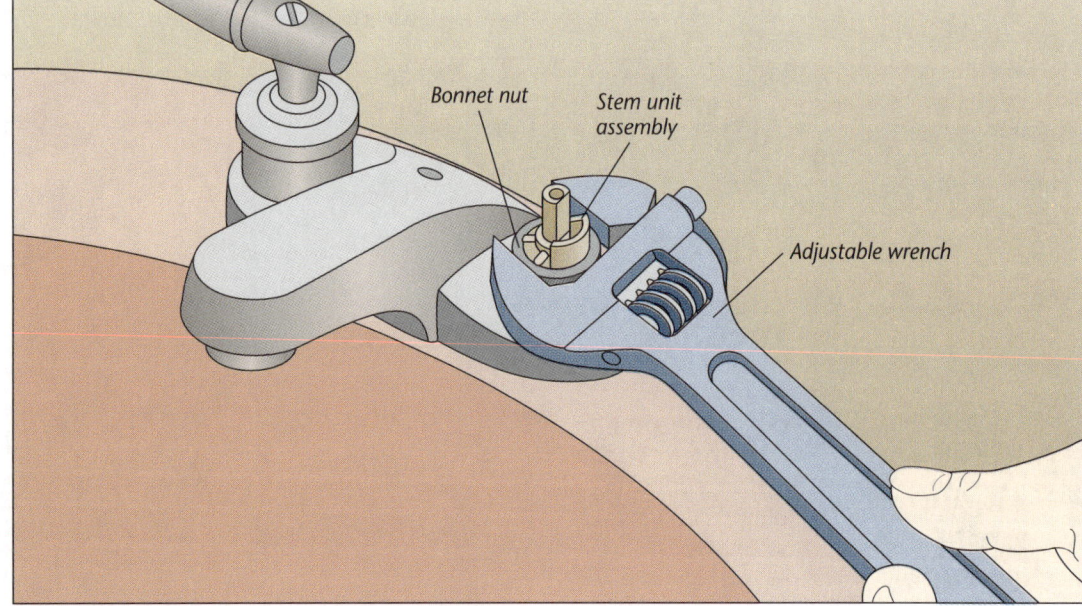

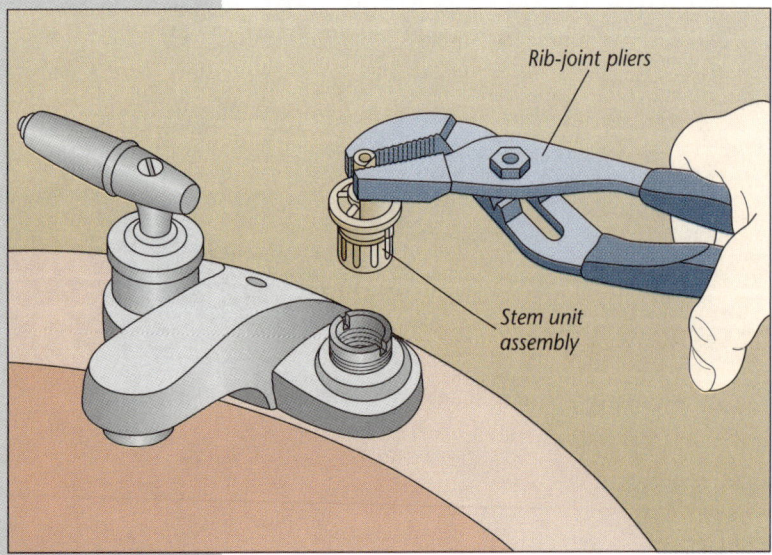

1 Disassembling the faucet
Pop off the trim cap, if there is one, using a blunt knife or screwdriver. Undo the handle screw and pull off the handle. Use an adjustable wrench to remove the bonnet nut *(above)*.

2 Replacing the O-ring or stem unit assembly
If the faucet is leaking from around the handle, the O-ring or stem unit assembly needs replacing. Pull out the stem unit assembly with rib-joint pliers *(left)*. If the O-ring is worn, replace it with an exact duplicate; lubricate the new ring with silicone grease before rolling it on. If the O-rings are in good condition, replace the stem unit assembly with a new one.

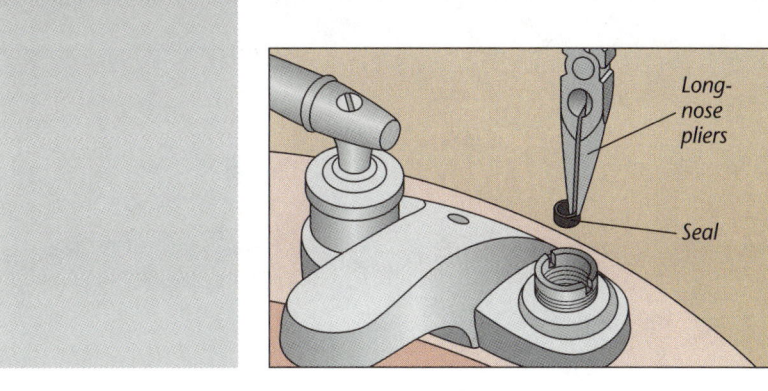

3 Replacing the seal and spring
If the faucet is dripping from the spout, the seal and spring probably need replacing. Pull them out with long-nose pliers and replace them with parts designed for the same model of faucet. To reassemble the faucet, put the stem unit assembly back. Be sure to line up the lugs in the assembly with the slots in the base of the faucet.

ROTATING-BALL FAUCETS

Inside every rotating-ball faucet is a slotted metal ball atop two spring-loaded rubber seals. Water flows when the openings in the rotating ball align with hot and cold water inlets in the faucet body.

If the handle of a rotating-ball faucet leaks, tighten the adjusting ring or replace the cam washer above the ball. If the spout of a ball faucet drips, the inlet seals or springs may be worn and need replacing. If the leak is under the spout, you must replace the O-rings or the ball itself.

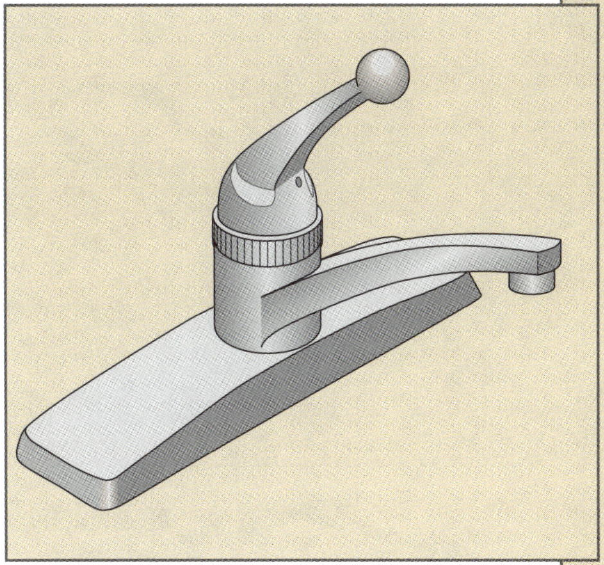

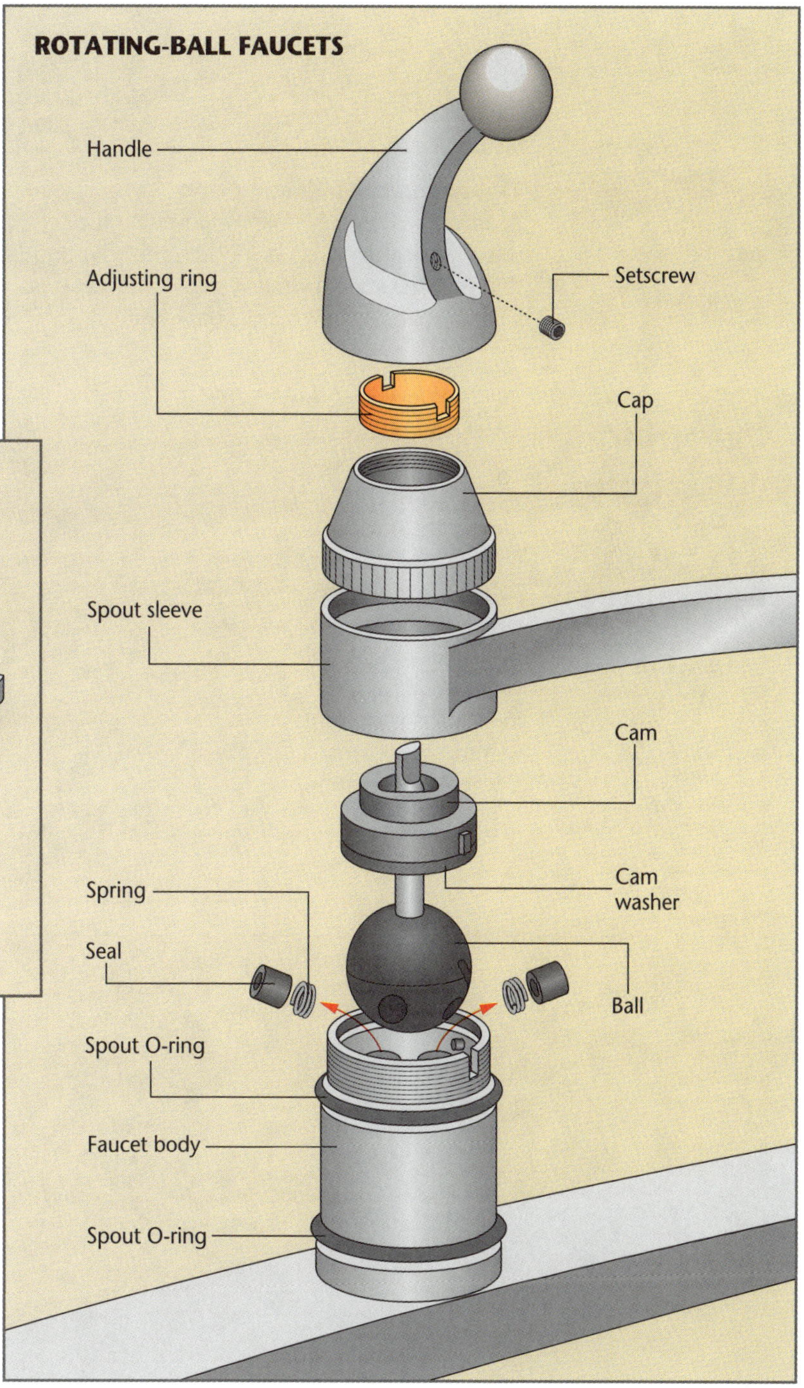

PLUMBING REPAIRS 39

Fixing a leaking handle on a rotating-ball faucet

TOOLKIT
- Hex wrench
- Adjusting-ring wrench or kitchen knife

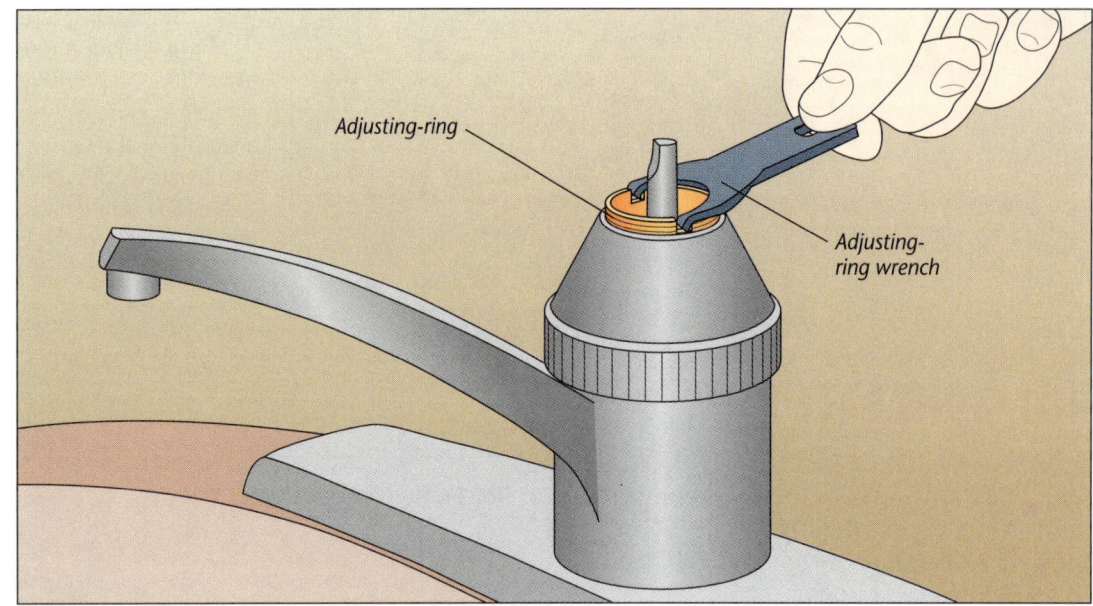

Tightening the adjusting ring
Remove the faucet handle by loosening the setscrew with a hex wrench. Tighten the adjusting ring with the special wrench shown *(above)* or a kitchen knife. Put the handle back on.

If this doesn't solve the problem, disassemble the faucet further, as explained in the repair for a dripping spout (below), and replace the cam washer above the ball.

Repairing a dripping spout on a rotating-ball faucet

TOOLKIT
- Hex wrench
- Rib-joint pliers
- Long-nose pliers
- Stiff brush or pocketknife

1. Disassembling the faucet
Loosen the setscrew with a hex wrench and remove the handle. Use tape-wrapped rib-joint pliers to unscrew the cap *(right)*. Lift out the ball-and-cam assembly. Underneath are two inlet seals on springs. Remove the spout sleeve to expose the faucet body.

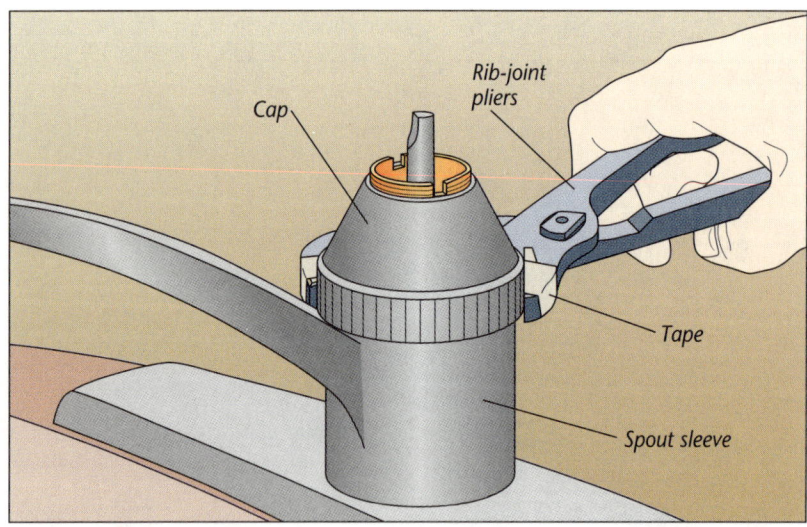

2. Replacing seals and springs
Use long-nose pliers to lift out the old parts *(left)*. With a stiff brush or pocketknife, remove any buildup in the inlet holes. If new spout O-rings are needed, apply a thin coat of silicone grease to them to stop leaks at the base of the faucet. Install a new spring and seal.

40 PLUMBING REPAIRS

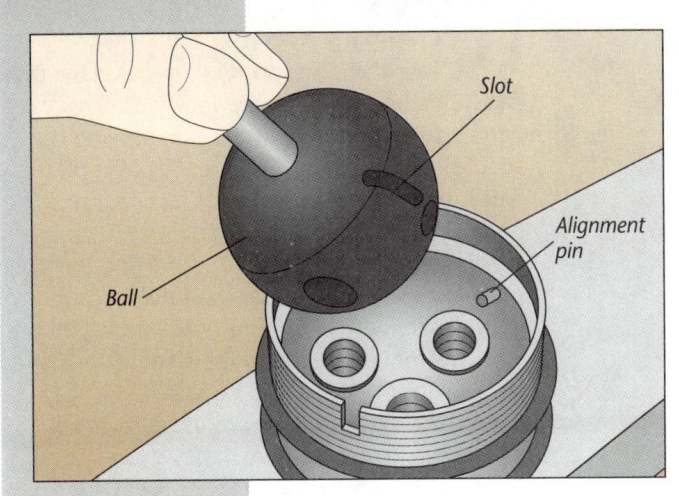
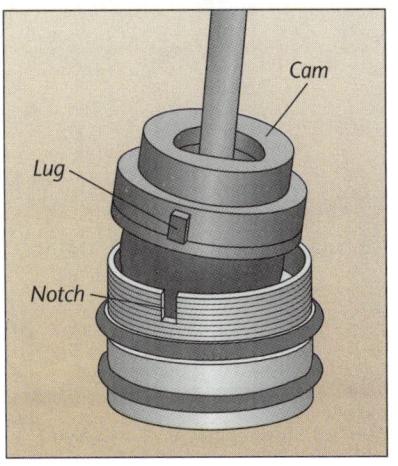

3. Reassembling the faucet

Before reassembling the faucet, check the ball; if it's corroded, replace it. To reinstall the ball-and-cam assembly, carefully line up the slot in the ball with the metal alignment pin in the faucet body *(far left)*. Also be sure to fit the lug on the cam into the notch in the faucet body *(left)*.

CARTRIDGE FAUCETS

These washerless faucets have a series of holes in the stem-and-cartridge assembly that align to control the mixture and flow of water. Usually, problems with this type of faucet occur because the O-rings or the cartridge itself must be replaced. If the faucet becomes hard to move, lubricating the cartridge O-rings should cure the problem.

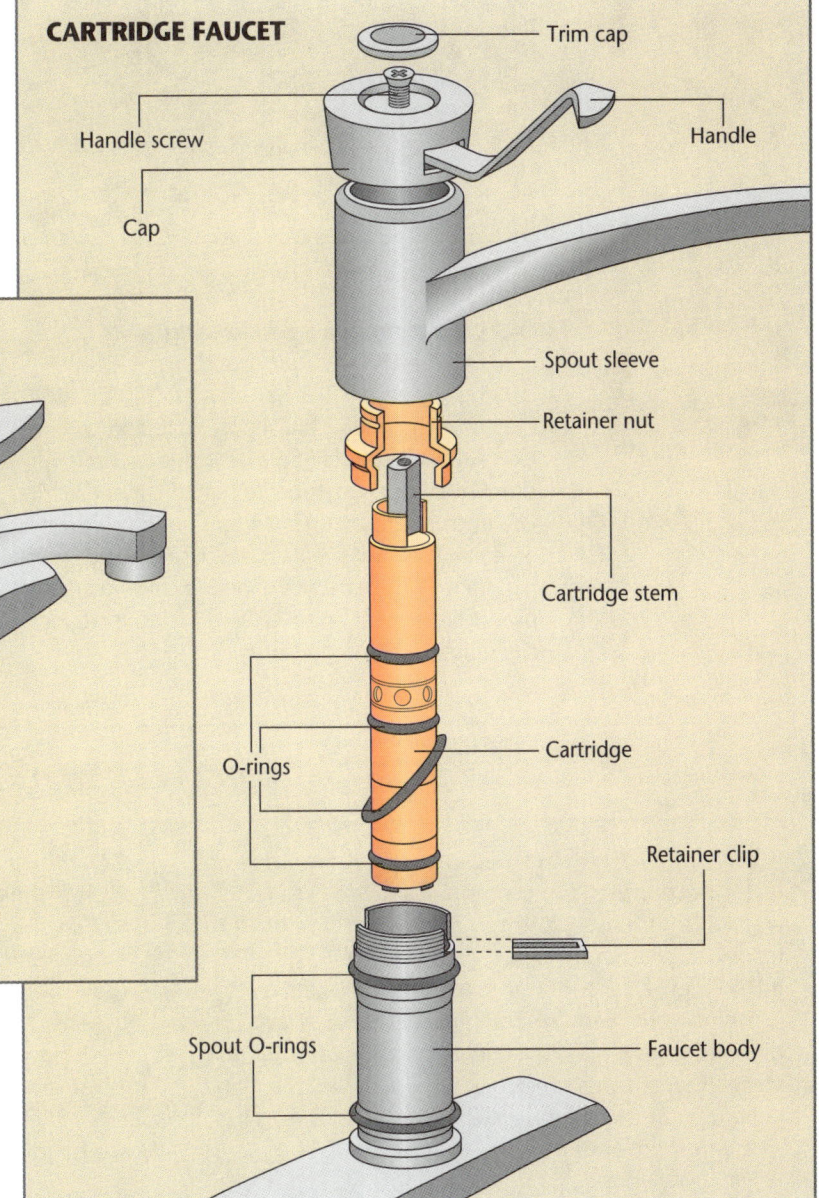

PLUMBING REPAIRS   41

Fixing a cartridge faucet

TOOLKIT
- Blunt knife (optional)
- Screwdriver
- Rib-joint pliers
- Long-nose pliers (optional)

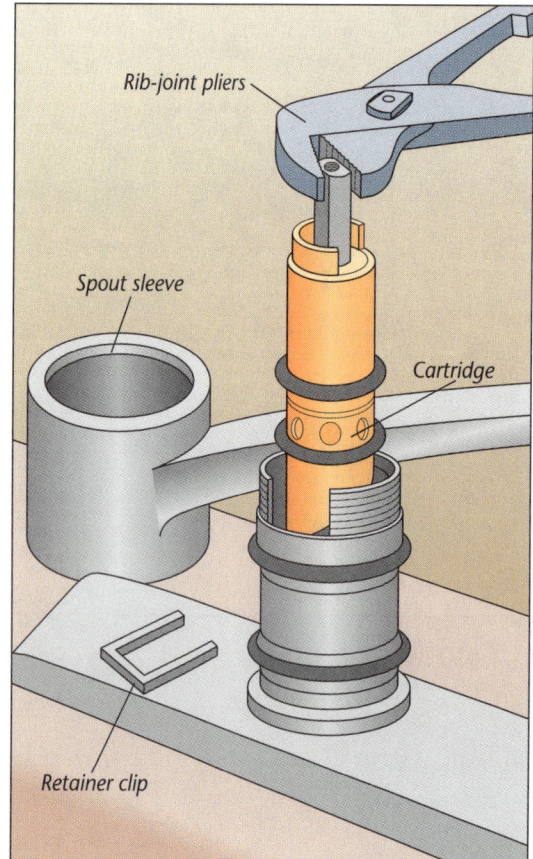

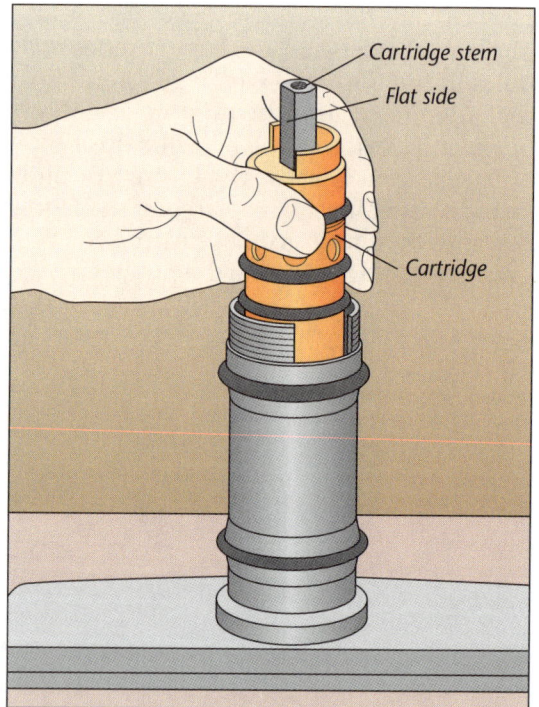

1. Changing an O-ring

Take the faucet apart by removing the trim cap with a blunt knife or screwdriver. Unscrew the handle screw and pull off the cap. Moving the spout sleeve back and forth, gently pull it off the faucet body. Then, unscrew the retainer nut using rib-joint pliers.

Next, remove the cartridge. You'll find the retainer clip just under the rim of the faucet body; using a screwdriver or long-nose pliers, pull the clip out from its slot. Grip the stem of the cartridge with pliers and lift it out *(left)*—it may require a strong pull. Examine the O-rings on the cartridge and replace them if they show signs of wear. Apply silicone grease to the new O-rings before installing them.

2. Putting in a new cartridge

If the O-rings are in good shape, it's the cartridge assembly that has seen its day. Take the old one to the plumbing supply store and buy an exact duplicate. Installing a cartridge *(right)* is a simple task, but remember to read the manufacturer's instructions first. Cartridges vary; the most common type has two flat sides, one of which must face front—otherwise, your hot and cold water supply will be reversed. Also, be sure to fit the retainer clip snugly back into its slot.

CONSERVING WATER

A kitchen or bathroom faucet uses about 5 gallons of water per minute. You can significantly reduce water wasted from these fixtures by following some or all of the following practices:

- Don't let water run; turn it off while you're shaving, brushing your teeth, or doing dishes. Rinse dishes for the dishwasher in a pan, not under running water, or just scrape them with a utensil, or wipe them with a cloth. Keep a basin in the kitchen sink; wash vegetables in it, cleaning them with a brush.
- Don't use running water to thaw frozen food.
- Chill drinking water by putting some in a container and storing it in the refrigerator.
- Attach a combination stream-spray aerator to the bathroom sink faucet if the spout will accommodate one. Use the steady stream to quickly wet a toothbrush; for handwashing, use the more efficient spray.
- Install a faucet shutoff at the end of the spout; you'll be more likely to use it than you are to turn off the water at the handle.

SINK SPRAYERS

Sink sprayers are handy for cleaning pots or the sink itself. They are attached to the faucet by a coupling nut under the sink. Sink sprayers have nozzle aerators that can clog, causing the diverter valve to malfunction. You can clean the aerator in the same way you would a faucet aerator—by disassembling it and cleaning the parts with a brush and soapy water. Use a pin or toothpick to clean out holes in the disc.

Leaks at the spray head can be solved with a new washer. Other leaks may require replacement of the hose. If problems persist, look to the diverter valve inside the base of the faucet. The diverter valve, which causes water to be rerouted from the spout to the sink sprayer hose, may need to be cleaned or replaced.

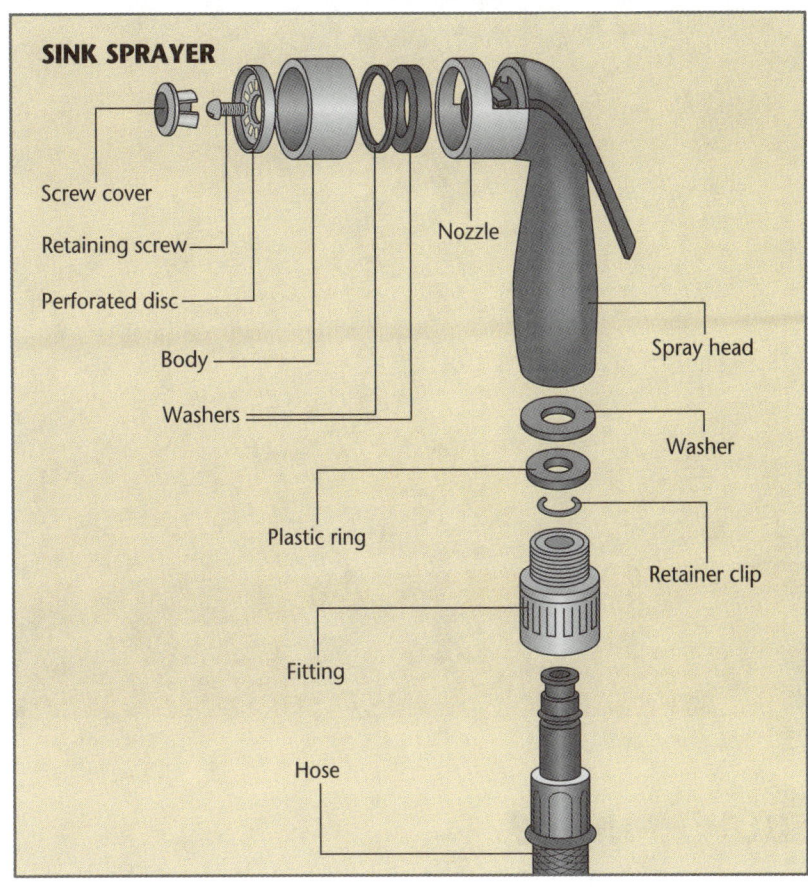

Servicing a sink sprayer

TOOLKIT
- Rib-joint pliers
- Small screwdriver
- Basin wrench (optional)

Stopping a leak
If the hose leaks at the spray head, try tightening the fitting at the base of the spray head. If this doesn't solve the problem, unscrew the head from the fitting. If the washer under the spray head is worn, replace it; then flush out the hose.

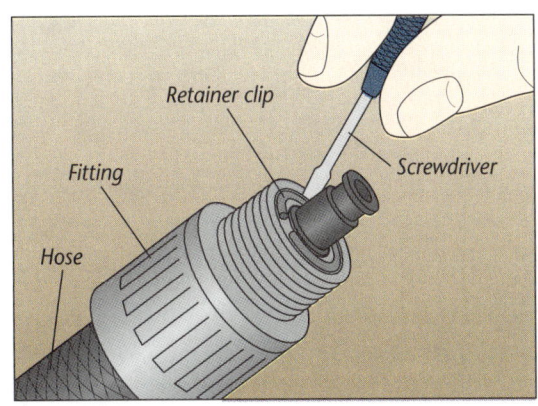

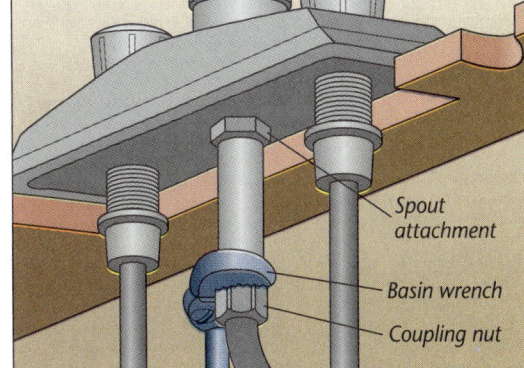

Replacing the hose
If the hose leaks where it attaches under the sink, undo the fitting from the tip of the hose by first removing the retainer clip *(above, left)*. Then, undo the coupling nut under the sink using rib-joint pliers or a basin wrench *(above, right)*—the hose might thread directly into the base of the faucet without a coupling nut.

Getting at the coupling nut can be awkward; you'll need to lie on your back under the sink. Once the coupling nut is detached, inspect the entire length of the hose for kinks or cracks. If you find defects, replace the hose with a new one of the same diameter—nylon-reinforced vinyl is the most durable.

PLUMBING REPAIRS

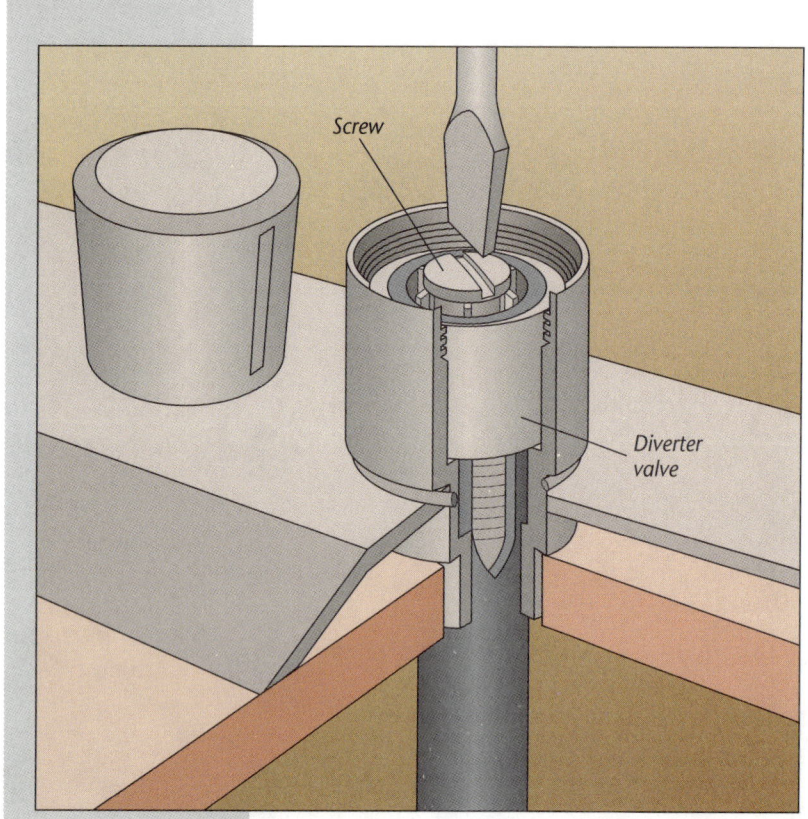

Cleaning a diverter valve

You'll need to take off the faucet spout to get at the diverter. Some spouts simply unscrew; for others, you may need to take apart the handle as shown on previous pages. Once you have access to the inside of the faucet body, loosen the screw atop the diverter valve *(left)* just enough to lift the valve from the seat. Take the valve apart and clean its outlets and surfaces with an old toothbrush and water. Soaking in vinegar or commercial lime-dissolver will help get rid of scale.

If cleaning the diverter doesn't improve the sink sprayer's performance, replace the diverter valve with an exact duplicate.

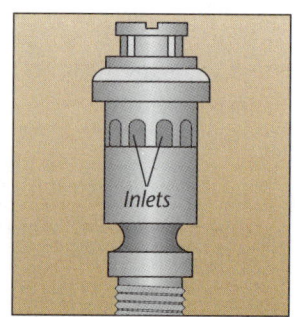

SHOWER HEADS

Before you decide to replace a leaking or water-wasting shower head, tighten all connections; if that doesn't stop the leak, replace the washer between the shower head and swivel ball.

If sluggish water flow is the problem, there's likely to be a blockage in the screen or face plate of the shower head. To cure this, remove the center screw and clean the face plate and screen in vinegar with a toothbrush.

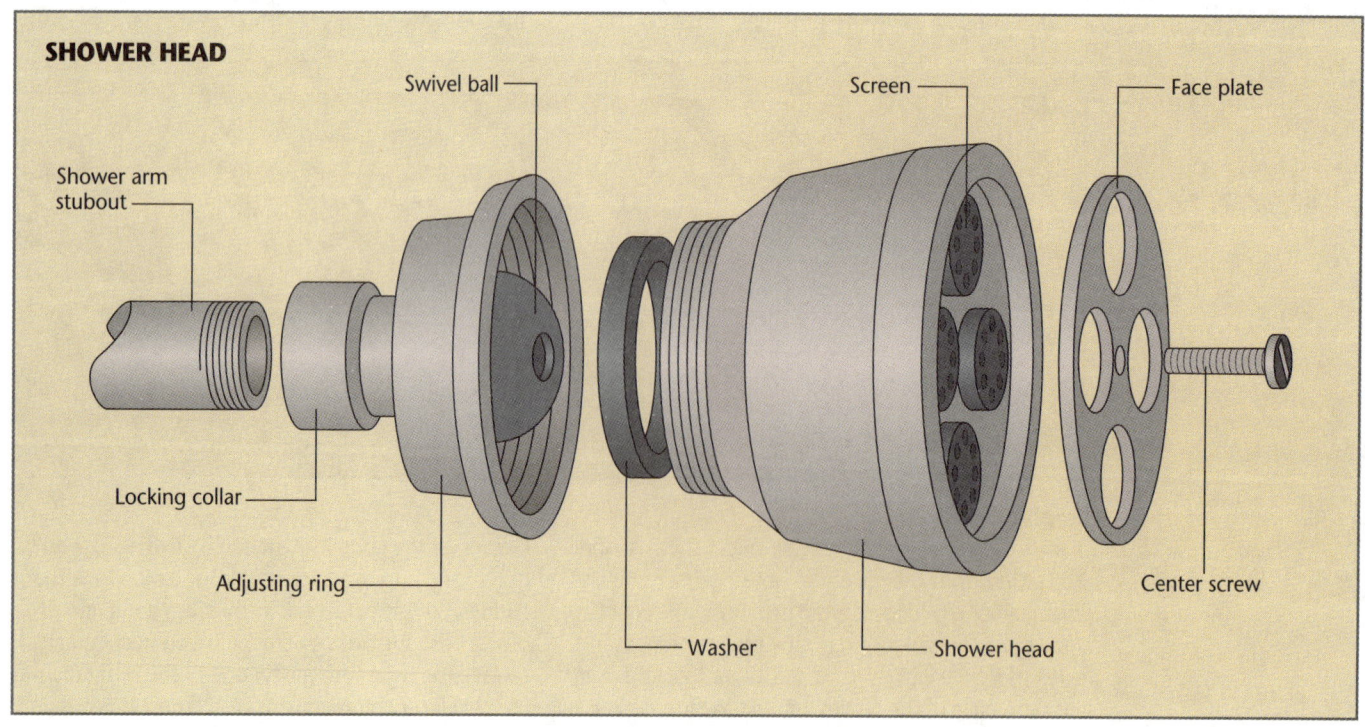

44 PLUMBING REPAIRS

SOLVING OTHER SINK PROBLEMS

If your sink doesn't hold water, or if there's a leak under the sink, it's time to examine the strainer—the part that opens and closes the drain and traps large food particles. This section will show you how to carry out basic repairs to two types of sink strainer.

The device that plugs sinks and tubs is now likely to be a pop-up assembly, not a rubber stopper on a chain. A defective pop-up may either open or close incompletely.

When working on either a strainer or pop-up, make sure that no one runs water into the sink or tub.

SINK STRAINERS

Modern stainless steel sinks sometimes have integral strainers which can't leak; however, more conventional strainers can occasionally give you problems. There are two types of sink strainers, as shown in the illustration at right. One is held in place by a locknut, the other by a retainer and three thumbscrews. Leaks can be caused by an improper seal between the strainer body and the sink. Check whether the strainer body is loose; if it isn't, the metal or fiber washer, or the rubber gasket between the locknut or retainer and the sink bottom, likely need to be replaced. This will require disassembling the strainer from under the sink.

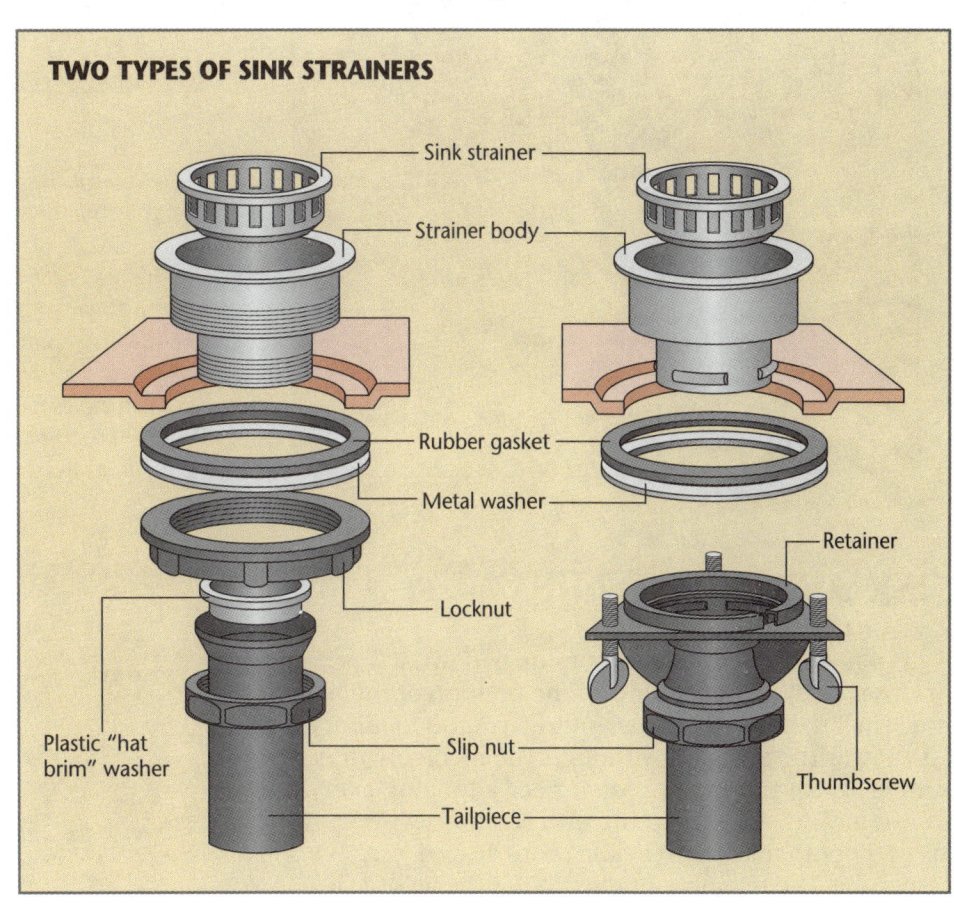

TWO TYPES OF SINK STRAINERS

Repairing a sink strainer

TOOLKIT
- Spud wrench
- Old screwdriver and hammer for locknut type (optional)
- Pliers and screwdriver for locknut type

1 ▶ Disassembling a strainer
Unscrew the slip nut with a spud wrench to free the tailpiece. For a locknut strainer, tap an old screwdriver with a hammer *(right)* to loosen the lugs on the locknut, or use a spud wrench. Be careful not to damage the sink. Remove the locknut, the washer, and the rubber gasket from the bottom of the strainer body. Lift the strainer from the sink. For a retainer-type strainer, simply undo the three thumbscrews on the retainer and disassemble in the same way as for a locknut type.

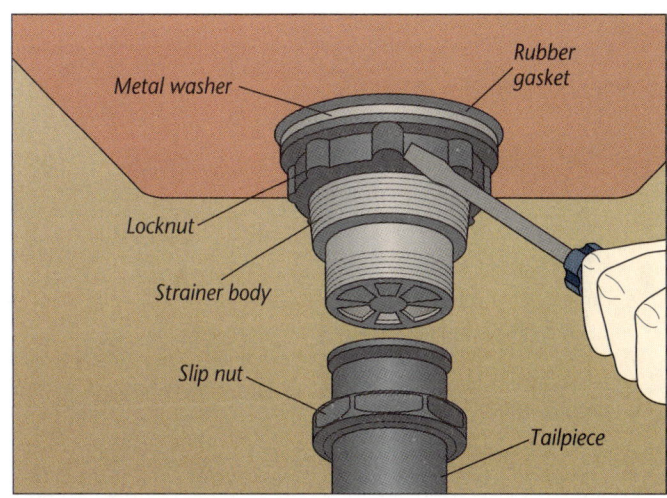

PLUMBING REPAIRS 45

2 **Sealing the drain opening**
Thoroughly clean the area around the drain opening. Check the rubber gasket and the metal or fiber washer for wear; get exact replacements if needed. Apply a 1/8" thick bead of plumber's putty around the underlip of the strainer body *(right)* and insert it in the sink opening. Press down firmly for a tight seal between the sink and the strainer.

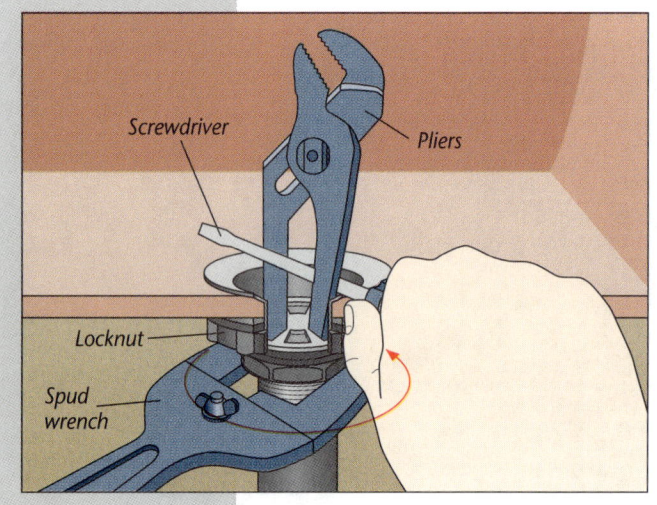

3 **Reassembling the strainer**
If the strainer is held in place by a locknut, work from beneath the sink to place the rubber gasket and metal washer onto the strainer body, and hand-tighten. Have a helper hold the strainer from above to prevent it from turning while you snug up the locknut with a spud wrench. If you're working alone, place the handles of a pair of pliers in the strainer and hold a screwdriver between the handles for counterforce while you tighten the nut *(left)*. Replace the slip nut and connect it to the tailpiece. Wipe any excess putty from the sink's surface.

SINK AND TUB POP-UPS

As its name implies, a pop-up pops up or down to open or close the drain depending on the position of the lift rod. The lift rod works through the pivot rod to raise and lower the stopper. Although the mechanism is simple, its several moving parts need adjusting every so often. Adjustments to the lift rod and clevis can cure a pop-up that doesn't close properly or doesn't open fully. The retaining nut that holds the pivot ball in place may need adjustment to prevent leaks.

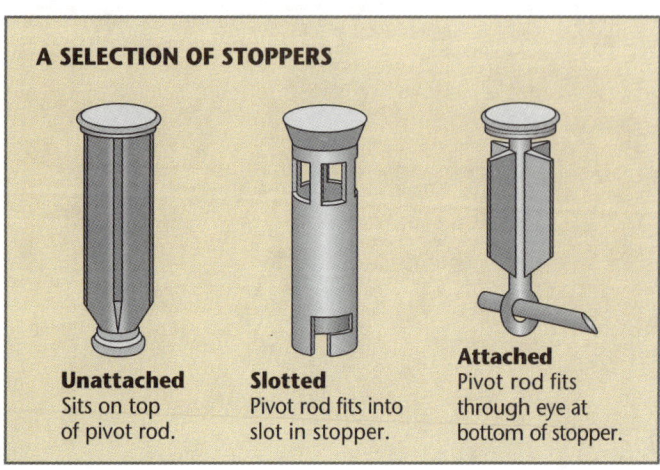

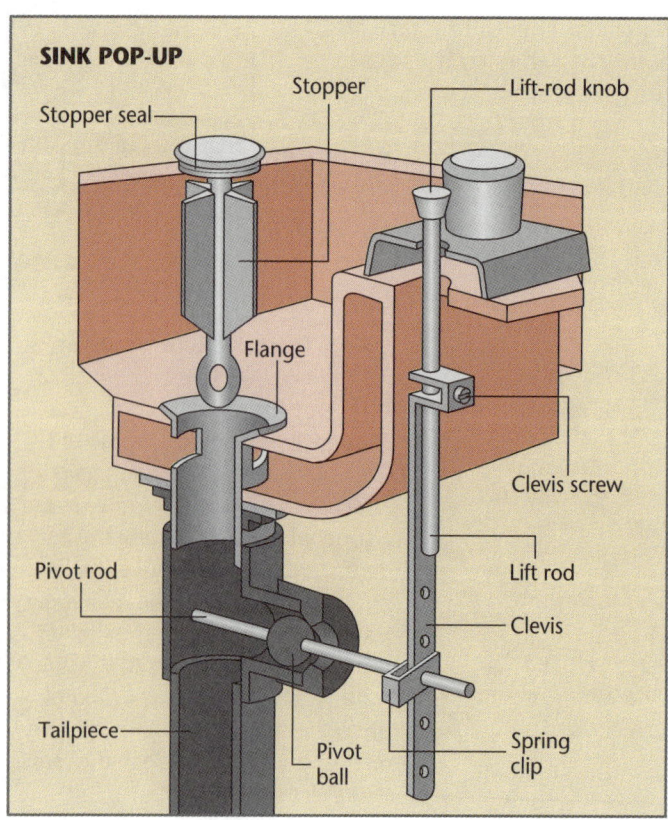

46 PLUMBING REPAIRS

Repairing a sink pop-up

TOOLKIT
- Screwdriver
- Rib-joint pliers

Working on the assembly

If the pop-up stopper isn't seating snugly, pull it out. The unattached type shown on the opposite page must be pulled out; the slotted type must be twisted to free it; for the attached type, you'll need to undo the retaining nut and pull out the pivot rod to remove the stopper. Clean the stopper of any hair or debris. Check its rubber seal; if it's damaged, pry it off and slip on a new one. If the pop-up still doesn't seat correctly, loosen the clevis screw, push the stopper down, and retighten the screw higher up. When the drain is closed, the pivot rod should slope downhill from clevis to drain body.

If the stopper is so tight that it doesn't open far enough for proper drainage, loosen the clevis screw and retighten it lower on the lift rod; if there is not enough room left on the rod, you'll need to move the spring clip and pivot rod to a lower hole. If water drips from around the pivot ball, try tightening the retaining nut that holds the ball in place. Still leaking? Replace the retaining nut and adjust the pivot rod so the pop-up seats properly.

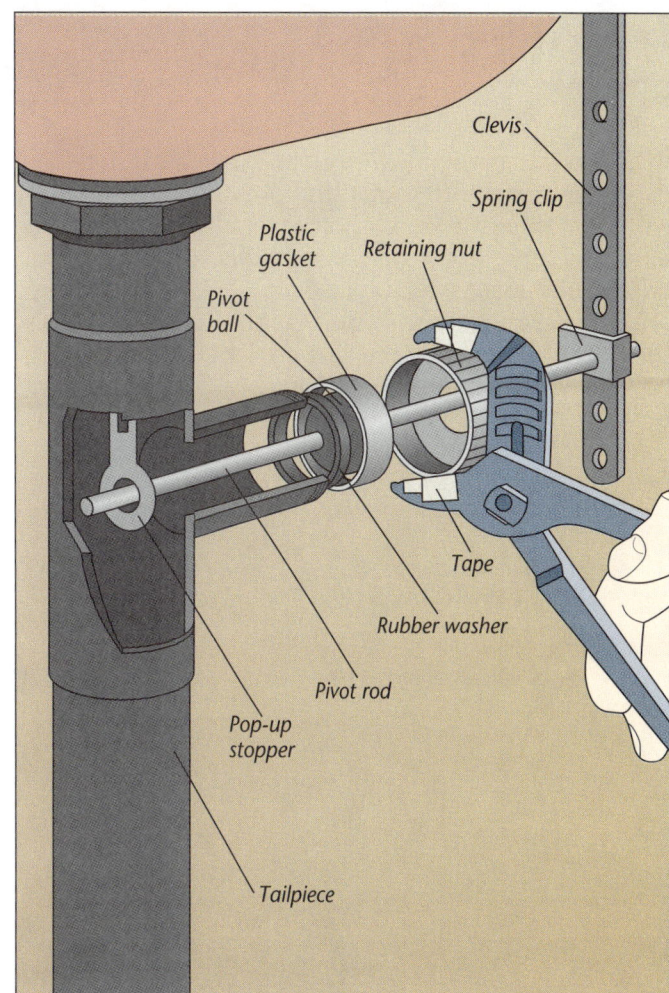

Servicing a tub pop-up

TOOLKIT
- Screwdriver
- Adjustable wrench

Adjusting the assembly

Remove the pop-up stopper and rocker arm, shown at right, by pulling the stopper straight up. Then unscrew and remove the tub's overflow plate and pull the entire assembly out through the overflow. If the stopper doesn't seat properly, loosen the adjusting nuts and slide the middle link up to shorten the striker rod. The striker spring rests unattached on top of the rocker arm.

For a sluggish drain, on the other hand, lower the link to lengthen the assembly. Before reassembling, clean the pop-up stopper.

NOTE: Instead of a pop-up assembly, some tubs have a strainer and internal plunger that blocks the back of the drain to stop the flow of water. The adjustments to the lift mechanism are identical.

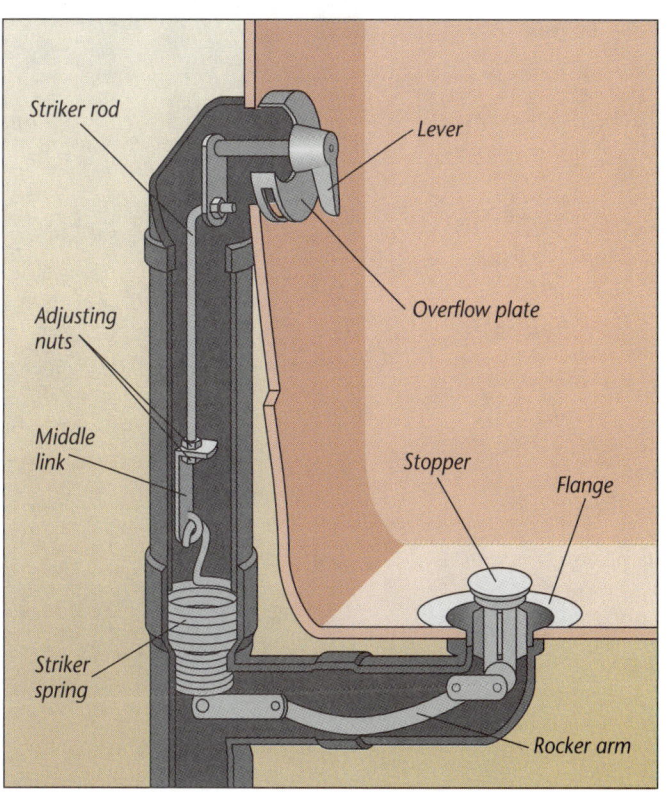

PLUMBING REPAIRS 47

TOILET REPAIRS

The workings of a flush toilet remain a mystery to most people until something goes awry. Fortunately, what may appear to be complex is, in fact, quite simple. Basically, there are two assemblies concealed under the lid: an inlet valve assembly, which regulates the filling of the tank, and a flush valve assembly, which controls the flow of water from the tank to the bowl. The toilet bowl includes a built-in trap. For a basic explanation of how your toilet works, see the opposite page. The illustration below and the chart on the opposite page should help you pinpoint a toilet problem.

If your toilet is whining or whistling, the inlet valve assembly may be to blame. The ball-cock type inlet valve, shown below, is the most conventional mechanism; both it and the diaphragm type use a float arm and ball. The float cup inlet valve *(inset)* eliminates the need for a float arm and ball. This device also helps prevent silent leaks. If the inlet valve is faulty, you may be able to just replace the washers or seal, or you may need to replace the whole assembly. Be sure any replacement assembly is designed to prevent backflow from the tank into the water supply.

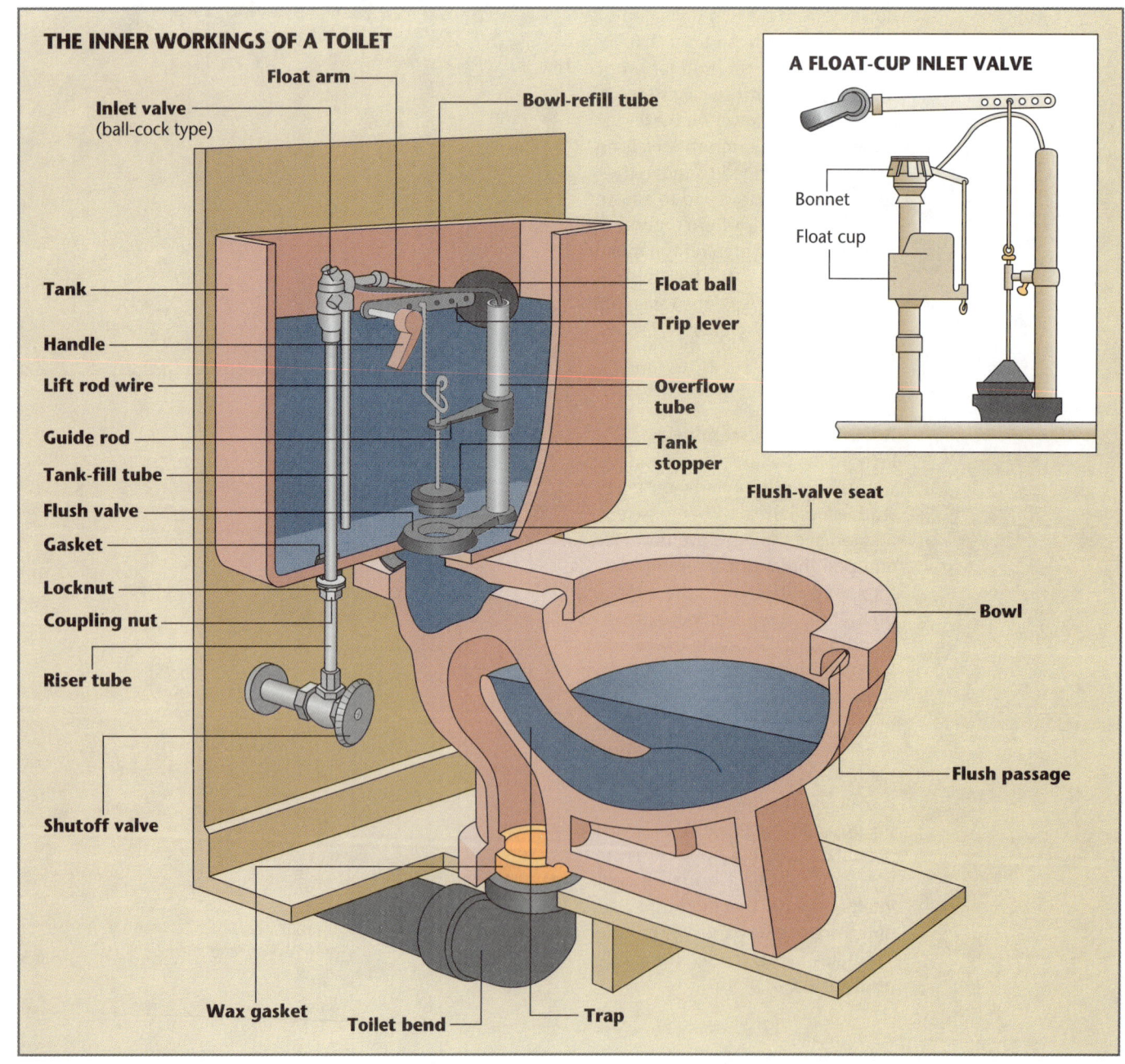

48   PLUMBING REPAIRS

If your toilet is constantly running, refer to the chart below for help in diagnosing the exact cause. Repairs include adjusting the float arm, or repairing or replacing the flush-valve assembly. Sometimes it's not obvious that your toilet is running. To detect the problem, see the tip on page 51.

The usual cause of a clogged toilet is an obstruction in the trap. To remove it, use a plunger; if that doesn't work, use a toilet auger. If these don't clear the clog, you might try using an auger or balloon bag in the nearest cleanout *(page 32)*.

Other miscellaneous toilet problems covered in this section are sweating (warm, moist air condensing on the tank), outright leaks, and difficulties with the handle.

For any toilet repair—unless you're simply adjusting the float arm—you'll need to shut off the water and empty the tank. You can shut off the water at the fixture shutoff or at the house shutoff valve *(page 10)*. Then, flush the toilet, holding the handle down, to empty the tank completely. Finally, sponge out any water that is left.

CAUTION: When you remove the tank lid, place it where it can't be hit by a falling wrench. Secondly, don't force a stubborn nut; oil it. This reduces the risk of slipping with the wrench and cracking the tank or bowl—a costly replacement.

HOW YOUR TOILET WORKS

Here's the chain of events that occurs when someone presses the flush handle on your toilet: The trip lever raises the lift rod wires (sometimes a chain) connected to the tank stopper. As the stopper goes up, water rushes through the flush-valve seat and down into the bowl via the flush passages. The water in the bowl yields to gravity and is siphoned out the built-in toilet trap to the drainpipe.

Once the tank empties, the stopper drops into the flush-valve seat. The float ball or cup trips the inlet-valve assembly to let fresh water into the tank through the tank-fill tube. While the tank is filling, the bowl-refill tube routes some water into the top of the overflow tube to replenish water in the bowl; this water seals the trap. As the water level in the tank rises, the float ball or cup rises until it gets high enough to shut off the flow of water, completing the process. If the water flowing into the tank fails to shut off, the overflow tube carries the excess water down into the bowl.

PINPOINTING THE PROBLEM

Problem	It may be...	Try this...
Noisy flush	Defective inlet valve	Oil trip lever, replace faulty washers, or install new inlet-valve assembly *(page 50)*
Continuously running water	Water level set too high	Bend float arm down or away from tank wall *(page 51)*
	Water-filled float ball	Unscrew ball and replace
	Tank stopper isn't seating properly	Adjust stopper guide rod or chain; replace defective stopper *(page 51)*
	Corroded flush-valve seat	Scour or replace valve seat *(page 51)*
	Cracked overflow tube	Replace overflow tube or install new flush-valve assembly *(page 52)*
	Inlet valve doesn't shut off	Oil trip lever, replace faulty washers, or install new inlet-valve assembly *(page 50)*
	Bowl-refill tube continues siphoning water into bowl after the tank is full	Position bowl-refill tube just inside top of overflow tube
Clogged toilet	Blockage in trap or drain	Remove blockage with plunger or toilet auger *(page 53)*
Inadequate flush	Faulty linkage between handle and trip lever	Tighten locknut on handle linkage, or replace handle *(page 56)*
	Tank stopper closed before tank empties	Adjust guide rod or chain so stopper rises 2" above valve seat with flush
	Water level set too low	Bend float arm up *(page 51)*
	Clogged flush passages	Poke obstructions from passages with wire
Sweating tank	Condensation	Install tank liner *(page 54)*
Leak between tank and bowl	Loose tank bolts or faulty spud washer	Tighten tank bolts, or replace spud washer *(page 55)*
Bowl doesn't refill	Bowl-refill tube is outside overflow tube	Position bowl-refill tube just inside top of overflow tube

PLUMBING REPAIRS

Repairing an inlet valve

TOOLKIT
- Rib-joint pliers
- Screwdriver

Disassembling and replacing worn parts

In a ball-cock assembly, worn washers may be the cause of loud tank noises. To replace them, remove the two thumbscrews on top of the ball-cock assembly that hold the float arm in place. Lift the float assembly out of the tank. With rib-joint pliers, pull the plunger up out of the ball cock. Inside the plunger is a seat washer and one or more split washers, as shown at right. If they're worn, replace them with exact duplicates.

In a diaphragm-type assembly, the bonnet (the top cover of the inlet valve) is screwed on; unscrew it and replace the washer or plunger inside. For a float-cup type, remove the bonnet by pushing down and turning clockwise; then replace the rubber seal inside.

If the inlet valve seems badly worn, or still leaks after the repairs, replace the entire assembly, as explained below.

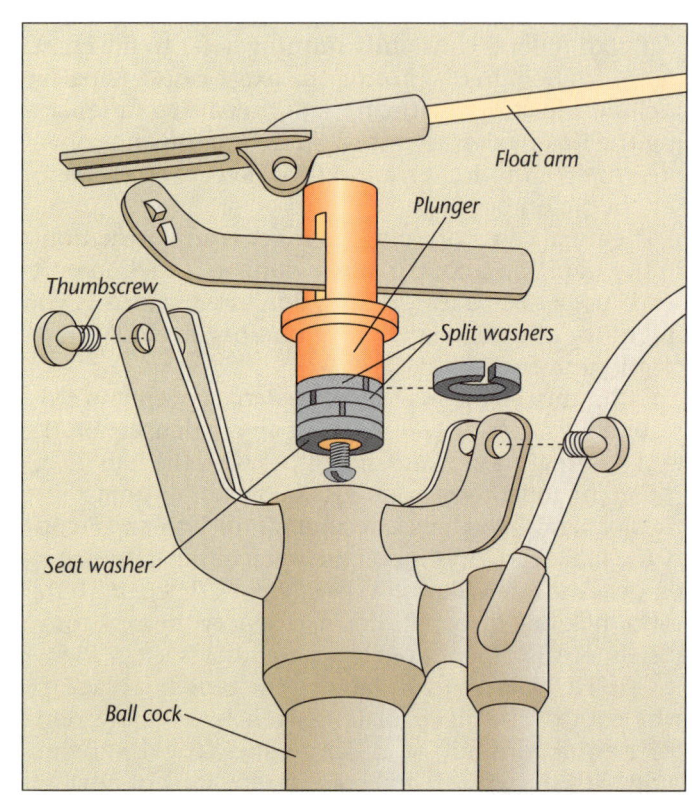

Replacing the inlet-valve assembly

TOOLKIT
- Adjustable wrench
- Locking pliers

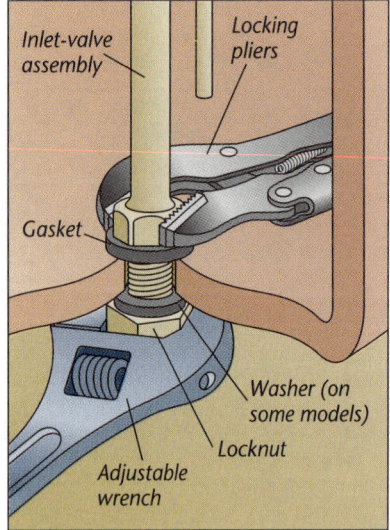

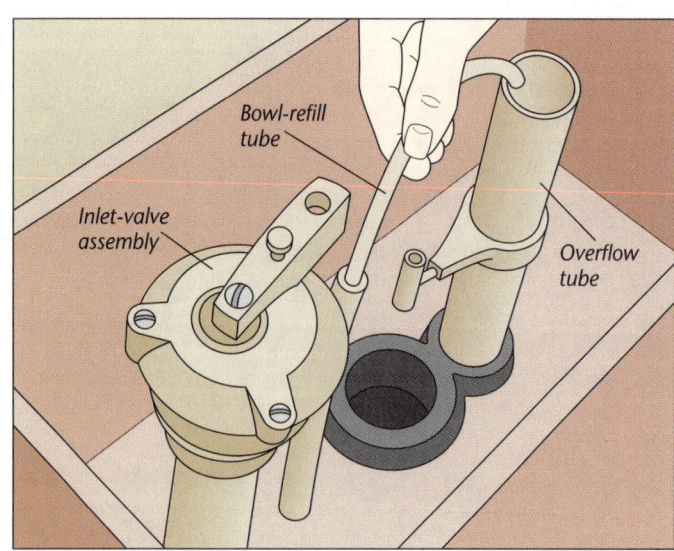

1 Removing the assembly
First, with an adjustable wrench, unfasten the coupling nut that connects the riser tube to the underside of the tank and remove the washer; examine the washer and replace it if it's worn. Remove the float ball and arm inside the tank. Then use a wrench to unfasten the locknut that holds the inlet-valve assembly to the tank (above)—hold on to the base of the assembly inside the tank with locking-grip pliers. Lift out the old inlet-valve assembly.

2 Installing a new inlet-valve assembly
To avoid any possibility of the new inlet valve siphoning water back into the water supply, buy one that is identified on the package as "anti-siphon" or "meets plumbing codes."

First, put on the inside gasket and install the new assembly in the tank. Then put the outside washer (if any) and locknut on the bottom of the new assembly. Tighten the locknut. Install the washer that goes inside the coupling nut and tighten the coupling nut on the riser tube under the tank. Place the bowl-refill tube just inside the top of the overflow tube; make sure the tip of the bowl-refill tube is not below the tank water level. Attach the float arm and ball and adjust them as described on the next page. Finally, turn on the water and check for leaks.

50 PLUMBING REPAIRS

MAINTENANCE TIP

DETECTING A RUNNING TOILET
A toilet that runs with a steady flow can waste as much as 7200 gallons of water a day. Of course, you can hear a toilet that runs this heavily. Harder to detect is a toilet that leaks slowly. Even an imperceptible toilet leak can waste as much as 40 gallons each day. Refer to the chart on page 49 for the likely causes of a running toilet.

To check your toilet for a leak, remove the tank lid and add about 12 drops of colored dye to the water inside. (Blue food coloring will work, or ask your water company for dye tablets.) Then, wait 5 minutes to see if the dye flows into the bowl. If it does, water is escaping either through the top of the overflow tube or past the tank stopper and valve seat.

Adjusting the water level

TOOLKIT
- Screwdriver for pressure-sensitive valve

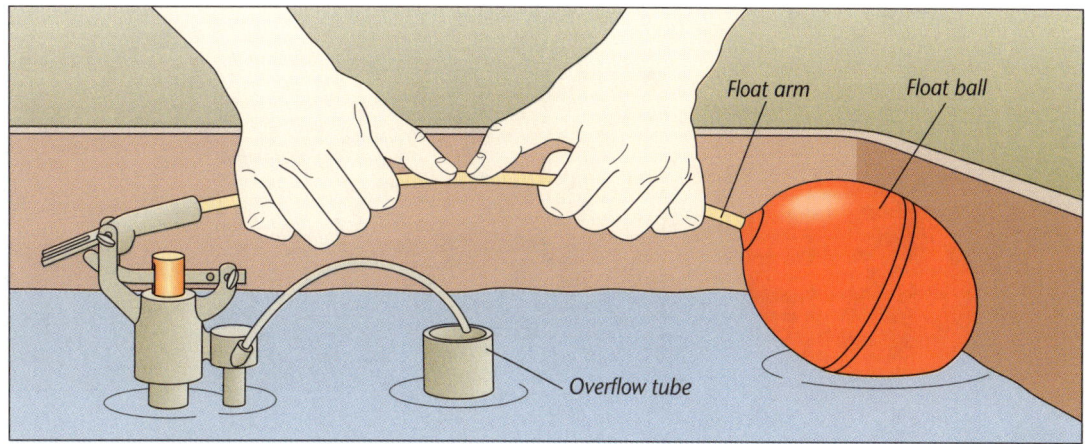

Adjusting the pressure-sensitive valve, float ball, or float cup

The water level should reach within 3/4" of the top of the overflow tube. If a waterline is printed inside the tank, use it as a guide.

If you have a unit with a float arm and ball, bend the float arm up to raise the level in the tank *(above)* or down to lower the level. Be sure to use both hands and work carefully to avoid straining the assembly. The float ball sometimes develops cracks or holes and fills with water; if this happens, unscrew the ball and replace it.

If you have a modern pressure-sensitive control, turn the adjusting screw clockwise to raise the water level, counterclockwise to lower it; one turn changes the level by about 1". If you have a float-cup valve, move the float up or down by squeezing the clip attaching it to the adjustment rod.

Repairing the stopper and valve seat

TOOLKIT
- Adjustable wrench

1 ▶ Inspecting and cleaning
A common cause of a continuously running toilet is a defective seal between the stopper and valve seat. To check, remove the lid and flush the toilet. Watch the stopper; it should fall straight down onto the seat. If it doesn't, make sure that the guide rod is centered over the flush valve.

Inspect the valve seat for corrosion or mineral buildup; gently scour with fine steel wool *(right)*. If the stopper is soft, encrusted, or out of shape, replace it *(step 2)*. If the water still runs after the stopper and valve seat have been serviced, replace the flush valve assembly *(page 52)*.

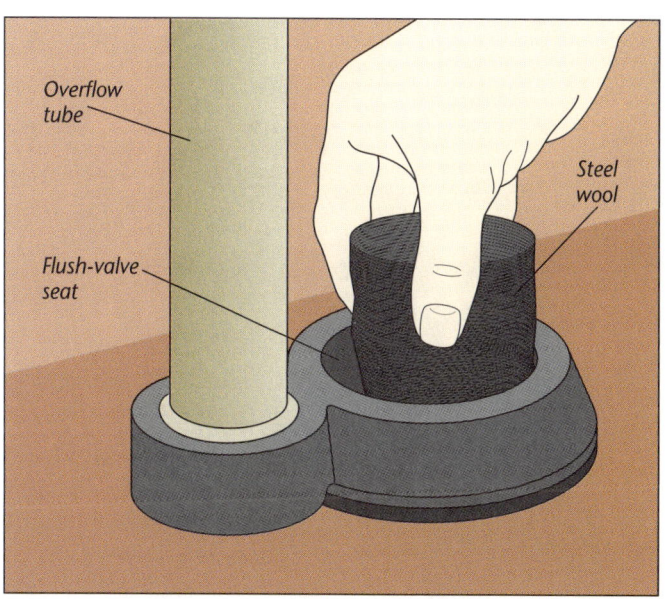

PLUMBING REPAIRS   51

2 **Replacing a stopper**
If the stopper needs replacing, install the flapper type with a chain *(inset)* to eliminate any future misalignment problems with the lift rod wires or guide arm, shown at right. Unhook the old lift wires from the trip lever, unscrew the guide rod, then lift out the guide rod and the wires. Slip the new flapper down over the collar of the overflow tube and fasten the chain to the trip lever. Adjust the chain with about 1 1/2" slack with the stopper in place on the flush valve.

Replacing a flush-valve assembly

TOOLKIT
- Spud wrench
- Adjustable wrench
- Screwdriver

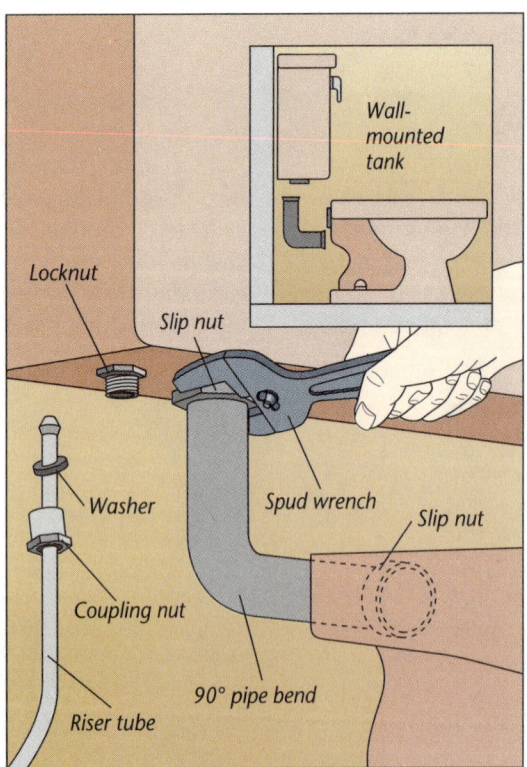

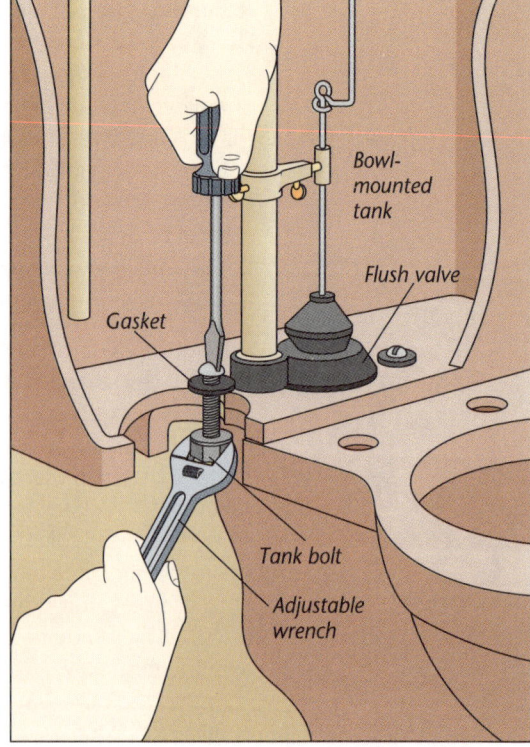

1 **Disassembling**
If you have an older toilet, with a separate wall-mounted tank, loosen the slip nut on the short 90° pipe bend under the tank using a spud wrench *(above, left)* and remove the pipe. For a bowl-mounted tank, use an adjustable wrench to remove the tank bolts *(above, right)*, then the gaskets; lift off the tank. Once you've removed the pipe bend or tank, unscrew the spud nut on the discharge tube (as seen in step 2) using a spud wrench; then remove the spud washer and flush valve.

52 PLUMBING REPAIRS

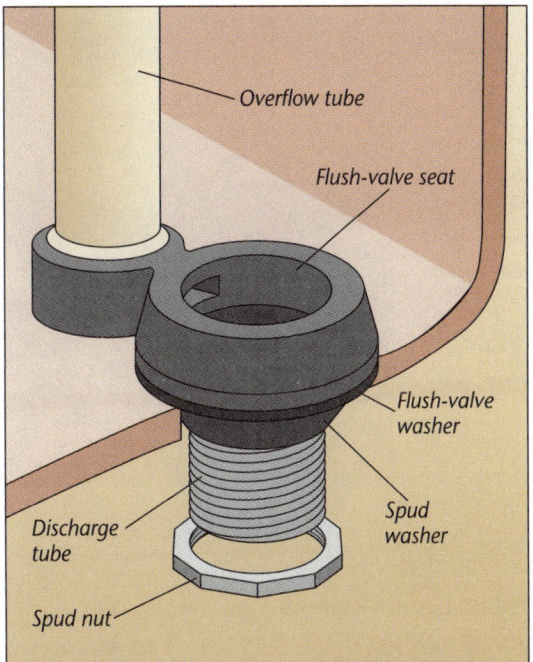

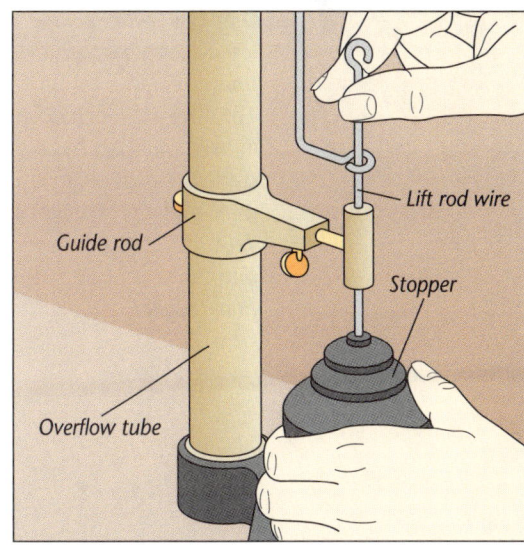

2 Installing a new assembly
Insert the discharge tube of the new valve assembly down through the tank bottom, with its washer against the tank *(above)*; then put on the spud washer from below and tighten the spud nut to hold it in place. Position the bowl-refill tube just inside the top of the overflow tube.

3 Installing the stopper
If your stopper has a guide rod and lift rod, center the guide rod on the overflow tube over the valve seat and tighten it in place. Install the lift rod wires through the guide rod and trip lever. Screw the stopper onto the lower lift rod wire *(above)*, aligning it with the center of the seat. For a flapper type, slip the flapper over the overflow tube, and fasten the chain to the lift rod with 1½" slack (with the stopper sitting on the flush valve). Turn on the water and check for leaks.

Unclogging a toilet

TOOLKIT
- Toilet plunger or toilet auger

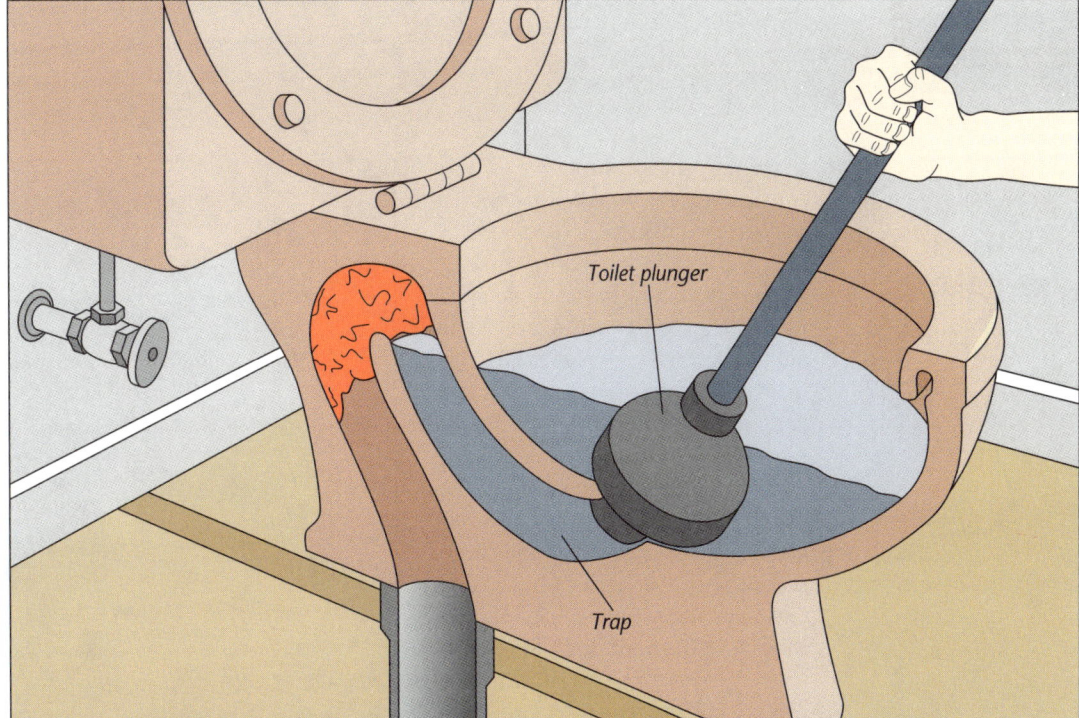

Using a toilet plunger
A toilet plunger has a special funnel-cup tip to fit the bowl, as shown above. To loosen a clog, pump the plunger up and down a dozen times and then pull it off sharply on the last stroke; the alternate pressure and suction should pull the obstruction back through the trap into the bowl.

PLUMBING REPAIRS

Using a toilet auger

If the plunger doesn't clear the clog, the next step is to use a toilet auger. This tool will reach down into the toilet trap *(right)*. It has a curved tip that starts the auger with a minimum of mess and a protective housing to prevent scratching the bowl. Follow the directions on page 29 for using a drain auger.

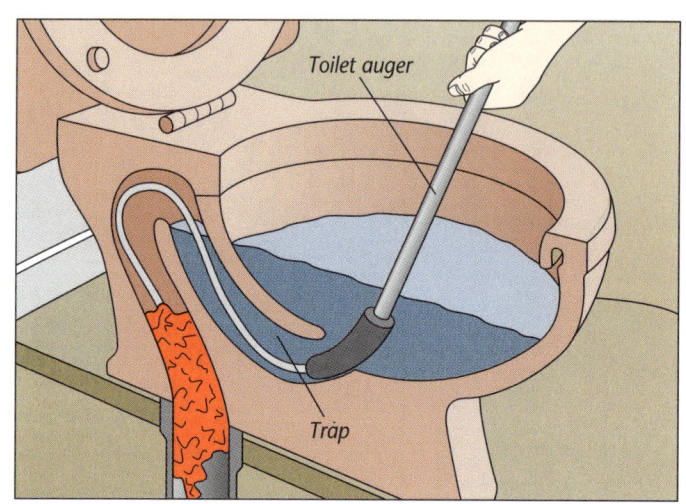

 QUICK FIX

PREVENTING A TOILET OVERFLOW
If you suspect a toilet is clogged, don't flush it or you'll have a flood on the floor. But if you see an overflow about to happen, you can usually prevent it by quickly removing the tank lid and closing the flush valve by hand; just push the stopper down into the valve seat.

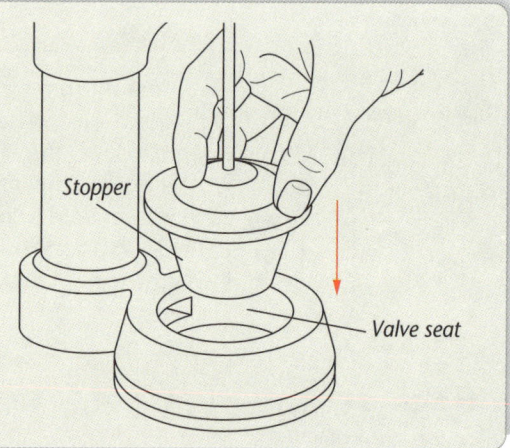

Preventing tanks from sweating

TOOLKIT
• Scissors or tin snips

Installing a tank liner

A sweating tank drips on the floor, encouraging mildew, loosening floor tiles, and rotting subflooring. An easy solution is to insulate the inside of the tank with a special liner sold at plumbing supply stores, or create a liner with foam rubber. To make your own tank liner, use scissors or tin snips to cut pieces of 1/2" thick closed-cell foam rubber to fit inside the tank.

To install the liner, apply a liberal coating of silicone rubber adhesive sealant or rubber cement to the dry tank sides and press the foam in place. Make sure the pads don't interfere with any moving parts. Let the adhesive dry for 24 hours before refilling the tank.

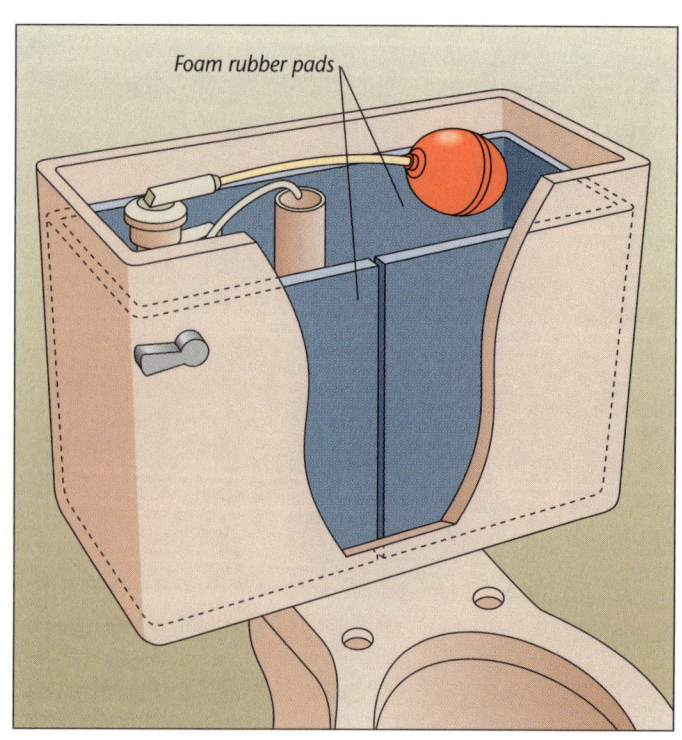

Stopping leaks between the tank and bowl or riser tube

TOOLKIT

For leaks between tank and bowl:
- Screwdriver and adjustable wrench

For leaks at top of riser tube:
- Adjustable wrench

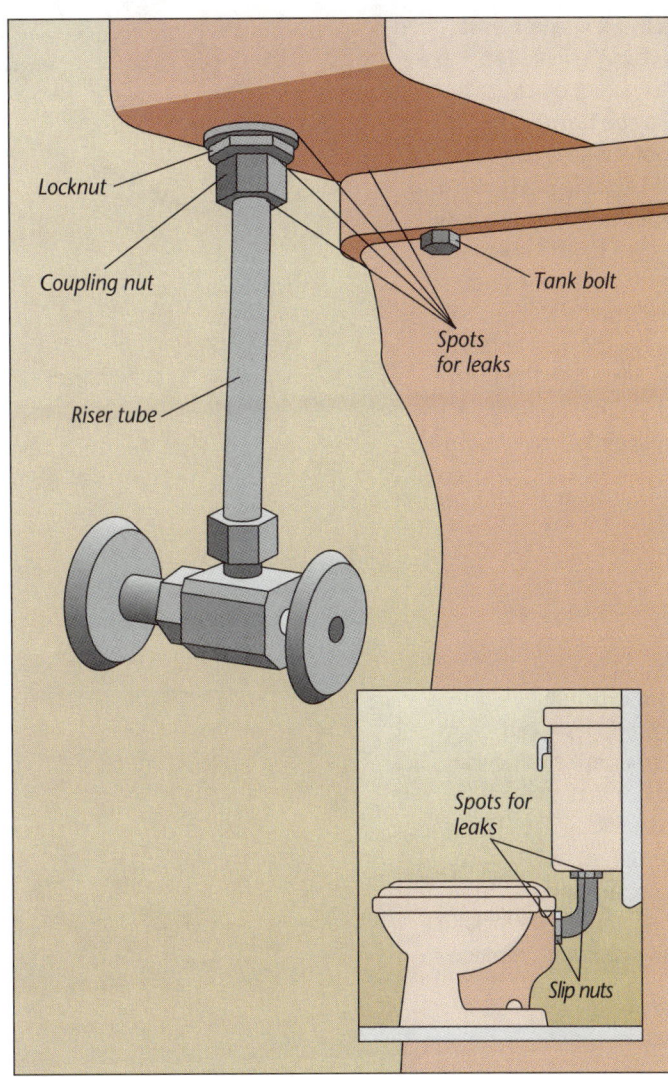

Tightening connections

To stop a leak between the tank and the bowl in a bowl-mounted tank toilet, try first tightening the tank bolts. Use a screwdriver on the inside of the tank and a wrench on the nut below the tank; be careful not to crack the porcelain. If this doesn't stop the leak, try replacing the tank bolts—they come with gasket, washer, and nut *(page 52, bottom right)*.

If there's still a leak when the toilet is flushed, remove the tank as explained on page 52. Replace the spud washer on the bottom of the flush valve *(page 53, step 2)*.

For a wall-mounted tank, try tightening the slip nuts connecting the 90° pipe bend to the tank and bowl; use a spud wrench. If this doesn't work, remove the pipe and replace the O-rings at either end.

If there is a leak between the tank and the locknut, tighten the locknut with an adjustable wrench. If this doesn't stop the leak, replace the gasket inside the tank *(page 50, bottom)*. If the leak is above or below the coupling nut at the top of the riser tube, try tightening the coupling nut. If the leak persists, disconnect the coupling nut and replace the washer hidden inside it.

Stopping leaks at the toilet's base

TOOLKIT
- Adjustable wrench
- Putty knife

1 ▸ Detaching the toilet and removing the gasket

If the base of the toilet bowl is leaking, first try tightening the bolts that hold it to the floor. If this doesn't stop the leak, you'll have to replace the wax gasket that seals the bowl to the floor. Begin by removing the tank, as explained on page 52. Then, remove the bowl, as shown on page 86. With a putty knife, remove the old gasket from the bottom of the bowl *(right)* and the floor flange.

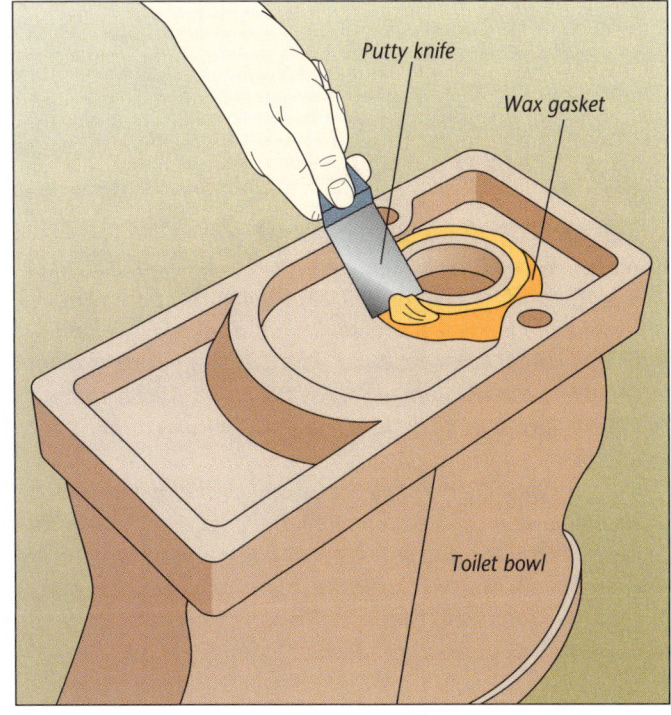

PLUMBING REPAIRS   55

2 **Replacing a gasket and reinstalling the toilet**

Place a new wax gasket on the toilet opening (called the horn). Check whether the floor flange is damaged and replace it if needed. This is also a good time to check whether the floor has been damaged by the leak and to repair it. Apply plumber's putty around the bowl's bottom edge.

Now replace the toilet by bolting down the bowl, installing the flush valve *(page 53)*, bolting on the tank, and reinstalling the inlet-valve assembly *(page 50)*.

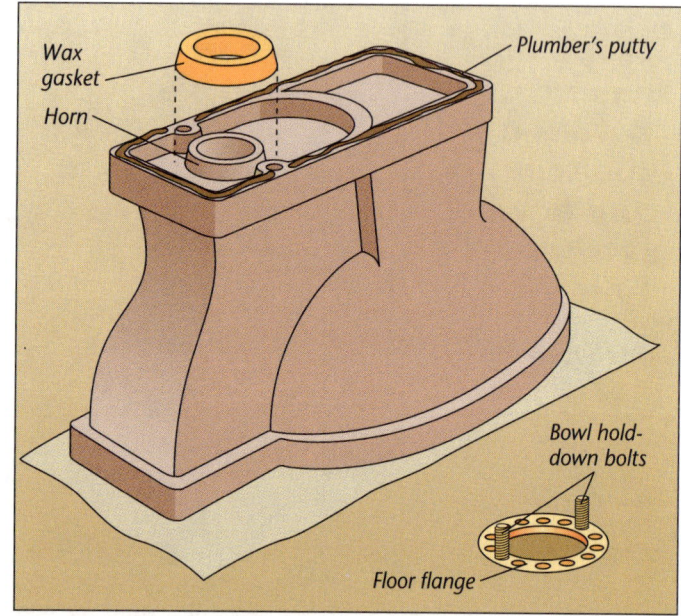

Fixing the handle

TOOLKIT
• Adjustable wrench

Tightening the locknut
A loose handle or trip lever can cause an inadequate or erratic flush cycle. Tighten the locknut that attaches the assembly to the tank. NOTE: This locknut tightens counterclockwise. Generally, the handle and trip lever are one unit; if tightening the locknut doesn't solve the problem, replace the entire unit.

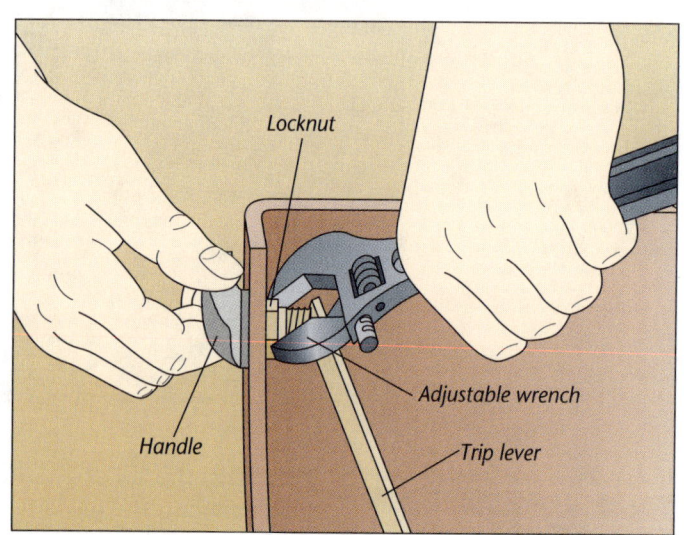

MAINTENANCE TIP

KEEPING UP YOUR SEPTIC TANK

A good septic tank system doesn't require a great deal of maintenance or call for many special precautions. But the maintenance it does require is crucial, since failure of the system can constitute a serious health hazard. You should have a diagram of your septic tank's layout, showing the location of the tank, pipes, manholes, and drainage field.

Chemicals, chemical cleaners, and thick paper products should never be disposed of through the system. Some chemicals destroy the bacteria necessary to attack and disintegrate solid wastes in the septic tank. Paper products can clog the main drain to the tank and smaller pipes to the dispersal field, making the entire system useless.

Have your septic tank checked at least once a year. To function properly, the tank must maintain a balance of sludge (solids remaining on the bottom), liquid, and scum (gas containing small solid particles). The proportion of the sludge and scum layers to the liquid layer determines whether pumping is needed.

Inspection and pumping should be done by professionals. Have your septic tank pumped whenever necessary, but plan ahead if you can. It's best to remove the sludge and a portion of the scum in the spring. If you have this done in the autumn, the tank will become loaded with solid waste that can't be broken down through the winter, when the bacterial action slows down.

PIPE PROBLEMS

A higher-than-normal water bill might give you the first indication of a leaking pipe. Or, you might hear the sound of running water even when all the fixtures in your home are turned off.

When you suspect that you have a leak, check first at the fixtures to make certain all the faucets are tightly closed. Then, go to the water meter *(page 7)*, if you have one. If the lowest-quantity dial on the meter is moving, it means that you're losing water somewhere in the system. If you don't have a water meter, you can buy a mechanic's stethoscope that amplifies sounds when it's held up to a pipe.

A faucet that refuses to run is the first sign of frozen pipes. In a severe cold snap, prevent freezing and subsequent bursting of pipes by following the suggestions on page 58. Even if pipes freeze, you can thaw them before they burst if you act quickly.

Pipe noises run up and down a non-musical scale, ranging from loud banging to high-pitched squeaking, irritating chatter, and resonant hammering. Listen carefully to your pipes; the noise itself will tell you what measures to take to quiet the plumbing *(page 59)*.

See page 69 for information on getting at pipes hidden in walls and ceilings.

Dealing with leaks

TOOLKIT
- Screwdriver for sleeve clamp
- Putty knife for patching with putty

1 Locating the leak

If you hear the sound of running water, follow it to its source. If you don't hear it, look for stains. If water stains the ceiling or is dripping down, the leak will probably be directly above. Occasionally, though, water may travel along a joist and then stain or drip at a point some distance from the actual leak. If water stains a wall, it means that there's a leak in a vertical section of pipe. The stain is most likely below the actual leak; you'll probably need to remove an entire vertical section of the wall *(page 69)* to find it.

If you don't hear the sound of running water, or see any telltale drips or stains, the leak is likely to be under the house in the crawl space or the basement; you should use a flashlight to check the pipes there.

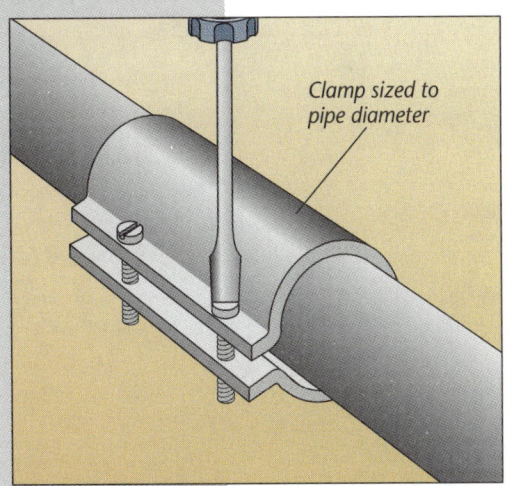
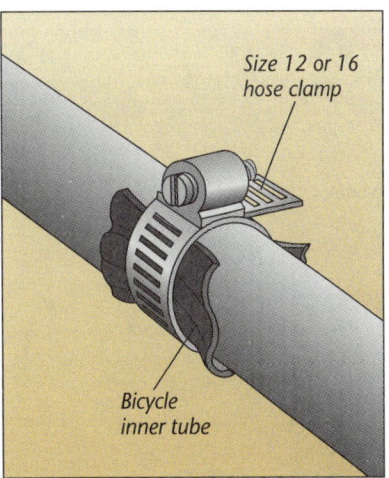
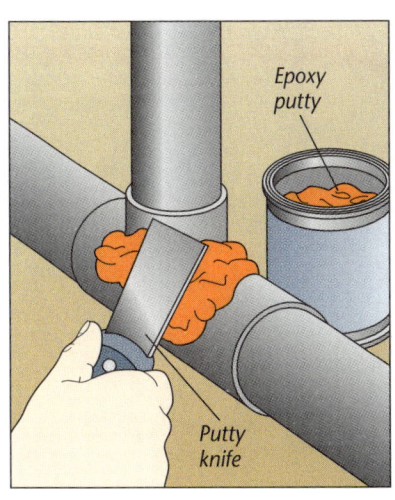

2 Fixing a leaking pipe

If the leak is major, turn off the water immediately, either at the fixture shutoff valve or the house shutoff valve *(page 10)*.

The ultimate solution is to replace the pipe *(pages 11 to 26)*, but there are simple temporary solutions until you have time for the replacement job. The methods described here for patching pipe are effective for small leaks only.

Sleeve clamps *(above, left)* should stop most leaks for some months, or even years; it's a good idea to keep some always on hand. Sleeve clamps are usually sold with a built-in gasket. A clamp that fits the pipe diameter exactly works best. An adjustable hose clamp *(above, middle)*, in size 12 or 16, stops a pinhole leak on an average-size pipe—be sure to use a piece of rubber such as a bicycle inner tube, or electrician's rubber tape with the hose clamp.

Epoxy putty *(above, right)* will often stop leaks around joints where clamps won't, but it doesn't hold as long. The pipe must be clean and dry for the putty to adhere; turn off the water supply to the leak to let the area dry.

PLUMBING REPAIRS   57

QUICK FIX

EMERGENCY PIPE REPAIRS USING COMMON SHOP SUPPLIES

If nothing else is at hand, you could use a C-clamp, a small block of wood, and a piece of rubber such as a bicycle inner tube. If you don't have a clamp, you can still stop a pinhole leak temporarily by plugging it with a pencil point—just put the point in the hole and break it off. Then wrap three layers of plastic electrician's tape extending three inches on either side of the leak. Overlap each turn of tape by half.

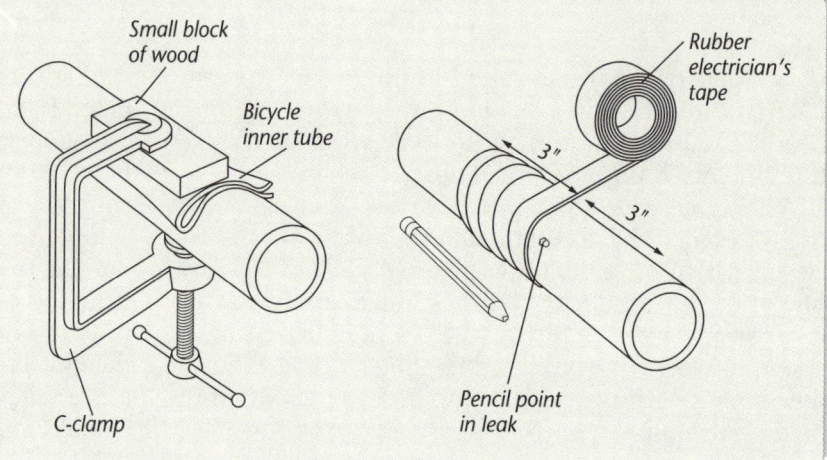

Thawing frozen pipes

TOOLKIT
- Hairdryer, heat gun, heating pad, or heat lamp

Warming the pipes

If a pipe freezes, first shut off the main water supply *(page 10)* and open the faucet nearest the frozen pipe. Waterproof the area with dropcloths, then use one of the following methods to warm the frozen pipe; be sure to work from the faucet back toward the iced-up area. When using an electrical device to thaw a pipe, make sure it is plugged into a GFCI outlet or extension cord to avoid the possibility of shock. Wear rubber gloves as backup protection, and avoid contact with water.

Use a hair dryer *(below, left)* or a heat gun turned on low to gently defrost the pipe. A heating pad *(below, middle)* is a gradual but effective method. For freezes that are concealed behind walls, floors, or ceilings, beam a heat lamp 8" or more from the wall surface. Be careful not to let water splash onto the hot glass. If no other method is available, wrap the pipe (except plastic) in rags and pour boiling water on it *(below, right)*; this method has the advantage that it doesn't involve electricity.

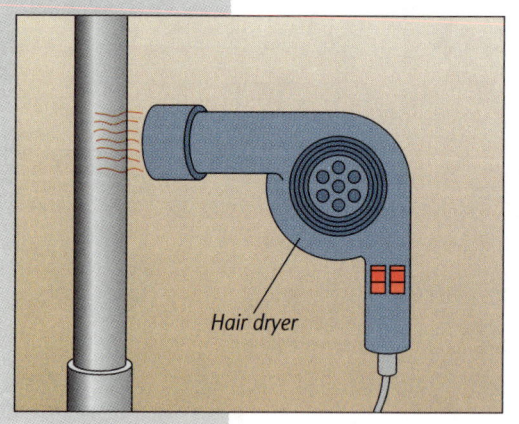
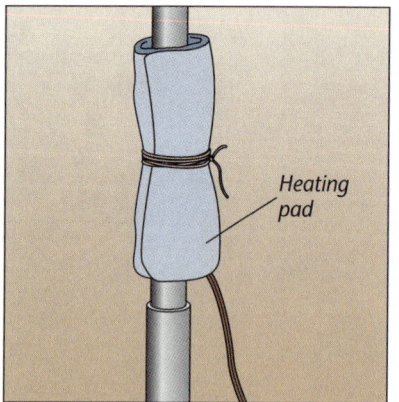
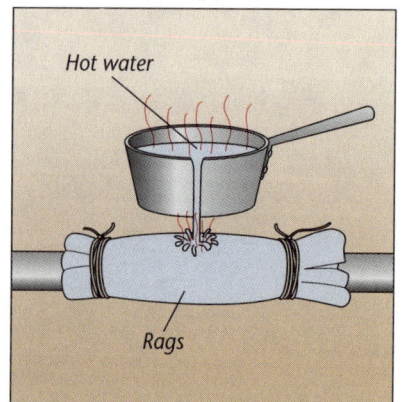

FORESTALLING FROZEN PIPES

Consider one of the following methods to stop your pipes from freezing: Keep a trickle of water running from faucets throughout the house. Aim a small lamp or heater at exposed pipes. Wrap uninsulated pipes with newspapers or foam. Install heat tapes (follow the manufacturer's instructions); insulation alone without heat tapes will not prevent freezing but it helps your heat tapes operate more efficiently. The best heat tapes have built-in thermostats that turn off when the outdoor temperature rises above freezing; this type should be used with plastic piping. Finally, keep doors ajar between heated and unheated rooms.

ASK A PRO

HOW DO I CLOSE DOWN MY PLUMBING FOR THE WINTER?
Homeowners who used to simply turn down the thermostat in a vacated house for the winter are now closing down the plumbing system because of prohibitively high energy costs. Winterizing your plumbing is a virtually cost-free alternative to frozen pipes.

First, turn off the house shutoff valve *(page 10)* or have the water company turn off service to the house. Starting at the top floor, open all faucets, inside and out. When the last of the water has dripped from the taps, open the drain plug at the house shutoff valve (you may have to contact the water company) and let it drain.

Turn off the energy to the water heater and open its drain valve. Empty water from the traps under sinks, tubs, and showers by opening cleanout plugs or, if necessary, removing, emptying, and reinstalling the traps *(page 61)*. Empty toilet bowls and tanks, then pour a 50/50 solution of automotive antifreeze and water into each toilet bowl and fixture.

If your home has a basement floor drain or a main house trap, fill each with full-strength automotive antifreeze/coolant, as shown at right.

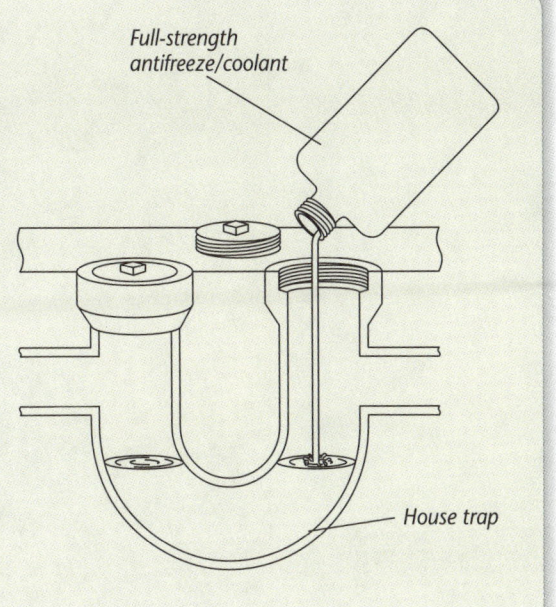

Quieting noisy pipes

Stopping water hammer
The most common pipe noise—water hammer—occurs when you quickly turn off the water at a faucet or an appliance. The water flowing through the pipes slams to a stop, causing a shock wave and a hammering noise.

Most water systems have short sections of straight or coiled pipe rising above each faucet or appliance, called air chambers. These sections hold air that cushions the shock when flowing water is stopped by a rapidly closing valve—the moving water rises in the pipe instead of banging to an abrupt stop *(below, left, and middle)*. These sections have a tendency to get completely filled with water and lose their effectiveness as cushions. To restore air chambers, take these steps: Check the toilet tank to be sure it is full; then close off the supply shutoff valve just below the tank. Close the house shutoff valve *(page 10)*. Open the highest and lowest faucets in the house to drain all water. Then, close the two faucets; reopen the house shutoff valve and the shutoff valve below the toilet tank. Normal water flow will reestablish itself for each faucet when you turn it on. (You can expect a few grumbles from the pipes before the first water arrives.)

If you're tired of restoring waterlogged air chambers, or if your house has none, install patented water-hammer arrestors *(below, right)*. These are designed to prevent waterlogging. Follow the manufacturer's instructions to install water-hammer arresters in both the hot and cold water lines as close to fixtures as possible.

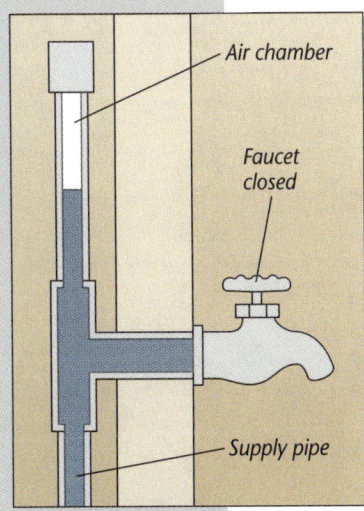 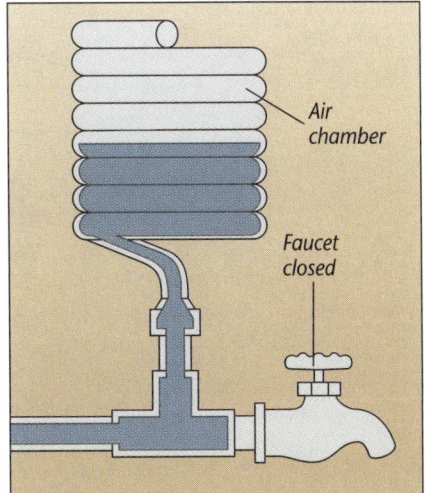 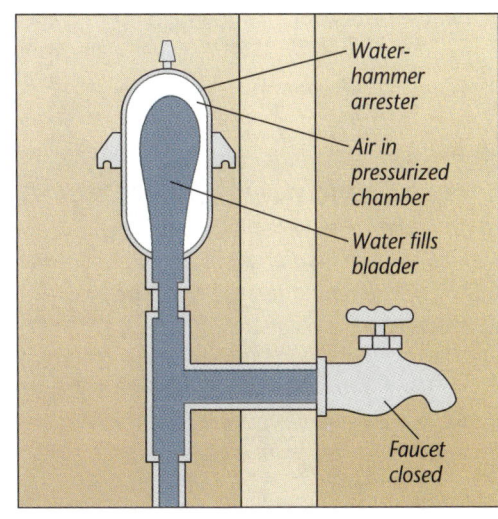

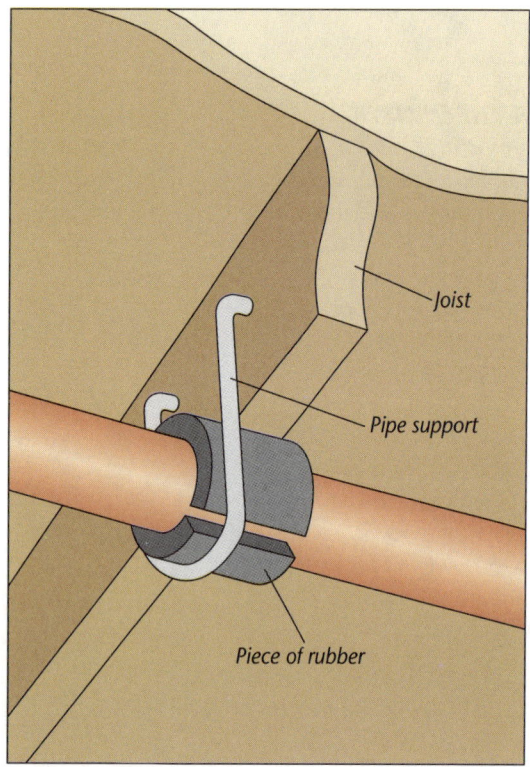 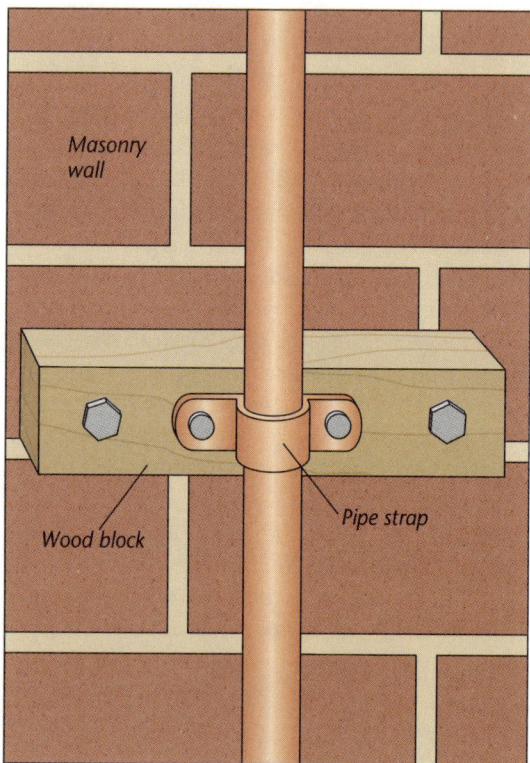

Eliminating banging, squeaking, or faucet chatter

Loud banging, chattering, or squeaking from your pipes indicates that wear or damage is taking place somewhere.

If you hear a banging noise whenever you turn on the water and it's not a water hammer problem, check the way the pipes are anchored. Banging pipes are generally easy to cure. You'll probably find the vibration-causing section of pipe is loose within its supports *(page 9)*. To eliminate banging completely, slit a piece of old hose or cut a patch of rubber and insert it in the hanger or strap as a cushion *(above, left)*. For masonry walls, attach a block of wood to the wall with masonry fasteners, then anchor the pipe to it with a pipe strap *(above, right)*. It's a good idea to install enough hangers to support the entire pipe run; if there are too few supports, the pipe will slap against the flooring, studs, and/or joists. Be careful not to anchor a pipe—especially a plastic one—too securely; leave room for expansion with temperature changes.

Pipes that squeak are always hot water pipes. As the pipe expands, it moves in the support, and friction causes the squeak. To silence this, insulate between the pipe and supports with a piece of rubber, as you would for banging.

Some noises may actually be caused by a fixture. Faucet chatter is the noise you hear when you partially open a compression faucet. To correct the problem, tighten or replace the seat washer on the bottom of the faucet stem *(page 35)* to prevent the stem from vibrating in the seat.

 QUICK FIX

CONTAINING OVERFLOW DAMAGE

To minimize water damage, follow these guidelines:
- Turn off the water supply at the house shutoff valve (page 10) *before taking time to trace the source.*
- If a pipe is leaking, spread waterproof dropcloths and pans.
- If a pipe has burst, dam doorways with rolled-up rugs or blankets. This will help to prevent water from spreading.
- Use a water-safe vacuum or rent a submersible-motor pump to remove large volumes of water.
- If an appliance is the culprit, turn off all the power by throwing the main circuit breaker or the main switch so that you're not working with electricity and water. CAUTION: Protect yourself from shock if the floor is damp by working with one hand and wearing thick rubber gloves, and by standing on a dry piece of wood or wearing dry rubber boots. If the floor is flooded, don't try to throw the breaker or main switch; instead call your utility company.

TRAP PROBLEMS

Traps are the workhorses of the drainage system. Traps remain filled with just enough water to keep sewage gases from coming up your drainpipe and entering the room. Unfortunately, by the nature of their shape and function, traps are often the first plumbing part to cause clogs or leaks.

If the trap of a sink, tub, or shower leaks, the cause could be corrosion or stripped fittings. If there's a clog and you have attempted to clean it out *(page 29)*, you can suspect a mineral buildup inside the trap. For any of these problems, the solution is to install a new trap and possibly a new fixture tailpiece.

Traps come in assorted sizes and shapes. Ordinarily, home plumbing systems use P-traps, either "fixed" or "swivel," shown at right. Although the S-trap shown on page 7 may be found in older homes, this type is no longer permitted by codes in many parts of the country.

Trap materials vary, as do sizes and shapes. Traps may be of brass, galvanized steel, or plastic. Longest-lasting are plastic and chrome-plated traps. The latter, however, are the most expensive of the lot.

Some traps may be attached to the tailpiece with a short extension with slip nuts. A trap adaptor is necessary to attach a metal trap to a plastic drainpipe.

Before doing any work, turn off the water at the fixture shutoff valves or the house shutoff valve *(page 10)*. Open a faucet at the low end to drain the pipes.

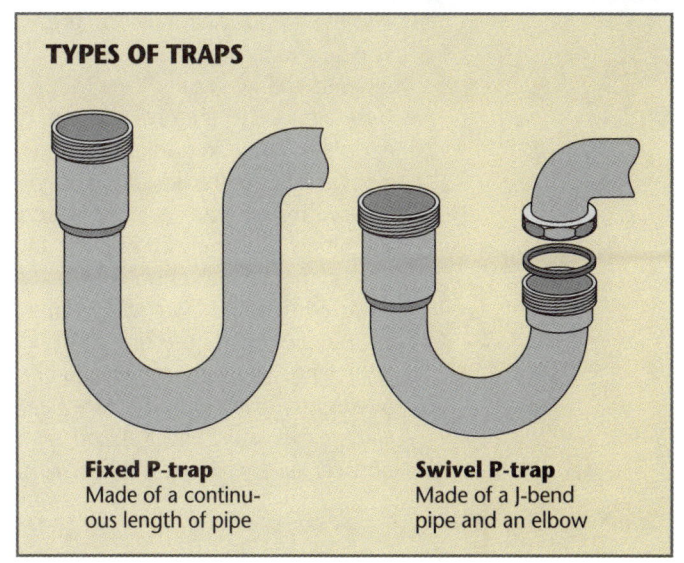

TYPES OF TRAPS

Fixed P-trap
Made of a continuous length of pipe

Swivel P-trap
Made of a J-bend pipe and an elbow

 **ASK A PRO**

DO I HAVE TO REPLACE MY S-TRAPS?
If you have S-traps in your home, you're not required to replace them with P-traps unless you're remodeling or altering part of your plumbing system.

Replacing a trap

TOOLKIT
- Adjustable wrench
- Spud wrench or rib-joint pliers

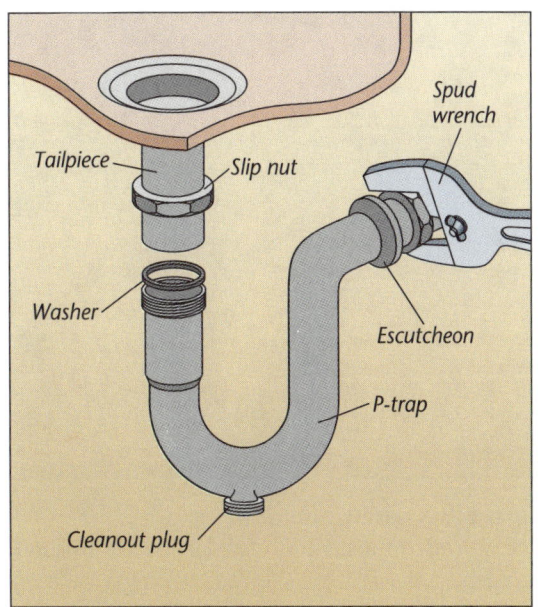
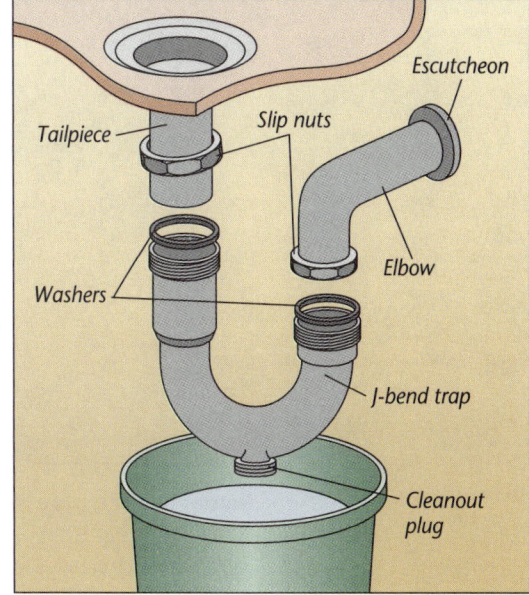

1 Removing the trap
The first step in removing the old trap is to remove the cleanout plug, if there is one, using an adjustable wrench (place a pail under the trap). Use a spud wrench (or taped rib-joint pliers for plastic) to loosen the slip nuts that attach a fixed P-trap to the tailpiece and drainpipe *(above, left)* and a swivel P-trap *(above, right)* to the elbow as well. Carefully pull off the trap.

PLUMBING REPAIRS 61

2. Attaching a new swivel P-trap

New traps are sold as complete units with washers, threaded slip nuts, and the fitting itself. Buy one the same size as the old one. Slide the new slip nuts and washers on over the fixture tailpiece in the order shown at right, connecting drainpipe and elbow. Coat the threads of the drainpipe with plumber's grease; attach the elbow. Set the trap in place and tighten the slip nuts at both ends by hand. With a spud wrench (or tape-wrapped pliers for plastic), finish tightening. Be careful not to strip or overtighten the slip nuts. Turn the water back on and check all connections for leaks while the fixture drains.

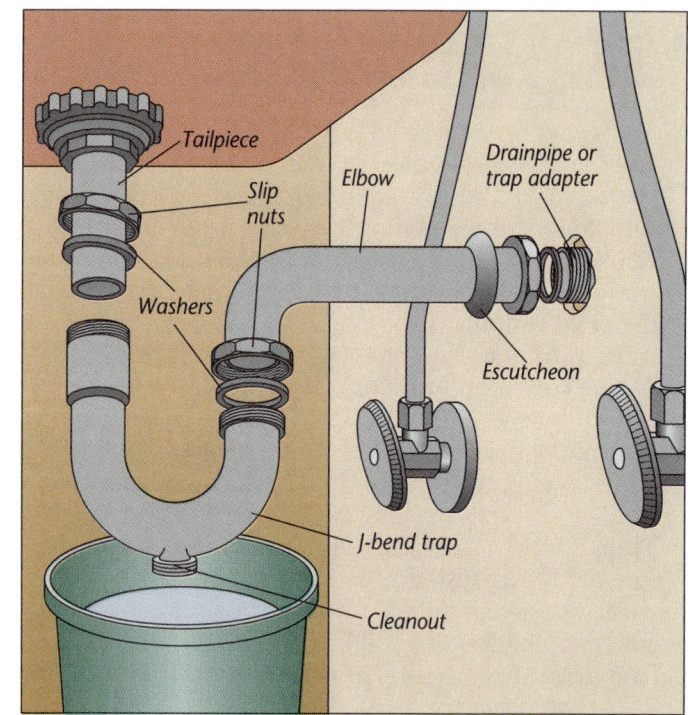

Replacing a tailpiece

TOOLKIT
- Spud wrench or rib-joint pliers

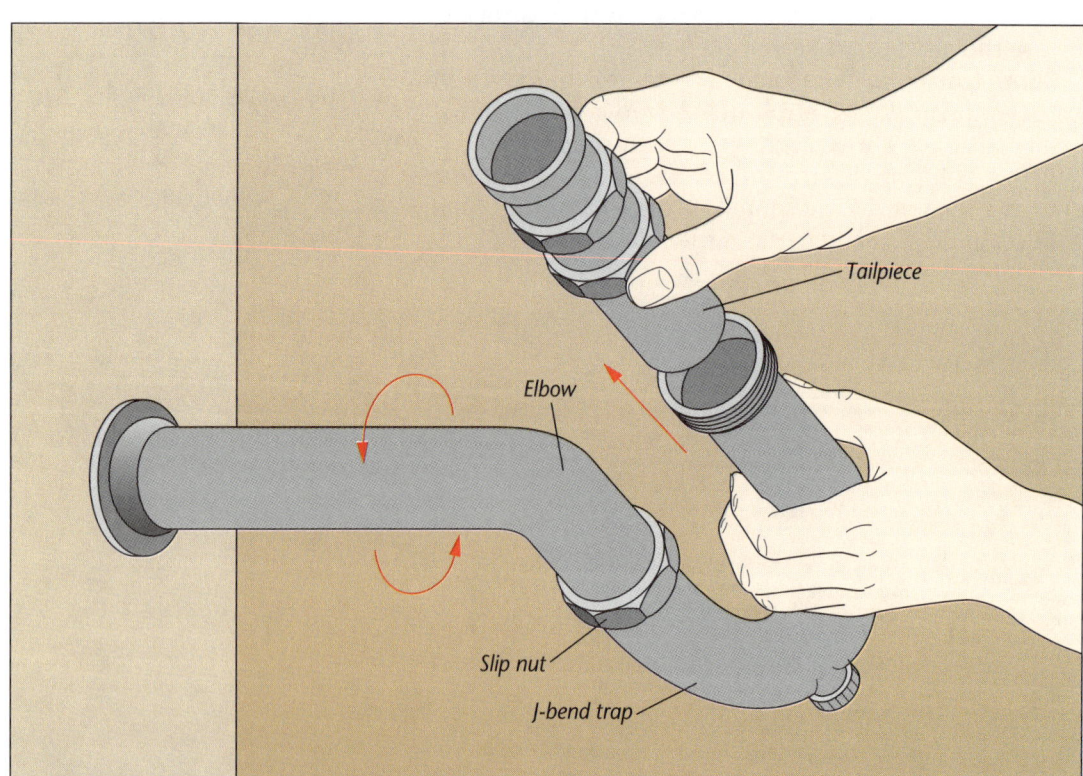

Removing the pipe and installing a new one

A tailpiece that is cracked or corroded must be replaced. Fortunately, replacements are available separately, and not just as part of a drain assembly.

To remove the tailpiece, unscrew the slip nuts that fasten it to the trap and the sink drain, and push the tailpiece down into the trap. Loosen the slip nuts at the drainpipe or elbow and turn the entire trap at the drainpipe a quarter turn *(above)*, just far enough to allow room to remove the tailpiece. You can now pull the old tailpiece out of the trap and replace it with a new one of the same length.

Coat all threads with pipe-joint compound to ensure a watertight seal. Hand-tighten all slip nuts, then use tape-wrapped rib-joint pliers or a spud wrench to snug them up. Restore water pressure, drain, and look for leaks.

STOPPING VALVE LEAKS

Valves, like faucets, control the flow of liquid through pipes by means of a simple mechanism: a handle that drives a stem down into its base to reduce or shut off flow. You'll find different types of valves for different uses: some of these restrict the flow even when fully open; others allow unrestricted flow when open.

The three most commonly used types are gate, globe, and angle valves.

Valves are available in a variety of materials. Those used in home plumbing are most commonly made of cast bronze, although plastic valves are sold for use with plastic pipe.

GATE VALVES

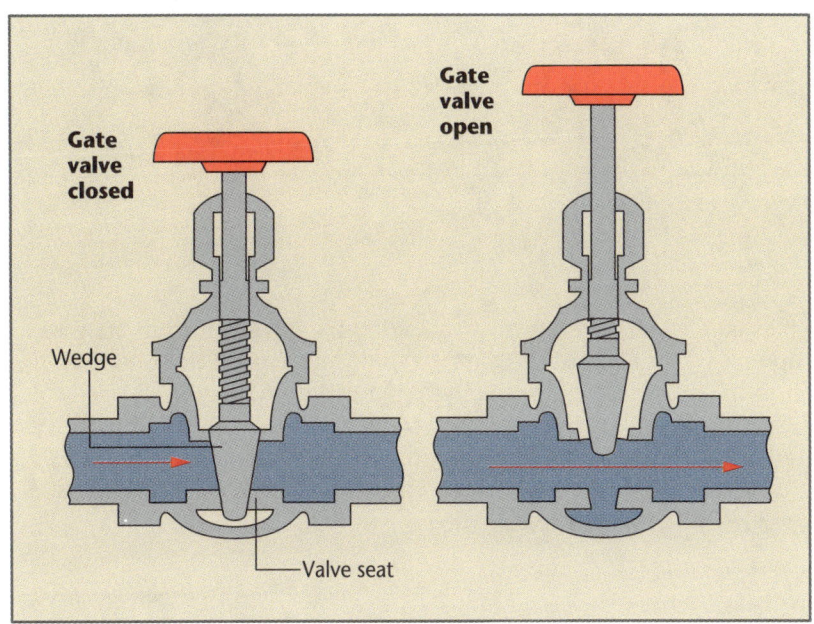

Used as the house shutoff valve in a residential system, a gate valve, shown at left, completely shuts off or completely opens up a supply pipe. This valve has a tapered wedge at the end of a stem that moves up or down across the water flow.

Because it takes a half dozen or more turns to fully open a gate valve, many people tend to open the valve only partially. Slightly opening a gate valve permits partial flow, but the pressure of the water moving across the wedge wears down its surface, resulting in an imperfect seal and a leaking valve. For this reason, gate valves should never be operated when only partially opened. If used correctly, these valves can last for many years.

GLOBE VALVES

Unlike a gate valve, a globe valve, shown at right, is designed to reduce water pressure: two half partitions change the direction or flow, slowing the water down even when fully open.

Like a compression faucet *(page 34)*, the globe valve has a stem that forces a washer into the valve seat. By turning the handle of a globe valve, you enlarge or diminish the opening for the water to pass through. Water supply pipe branches are usually equipped with globe valves, which can withstand frequent opening and closing under pressure.

This valve is the easiest kind to repair. Most often, the seat washer (or, sometimes, a disc) is defective and needs replacement; see the next page for instructions.

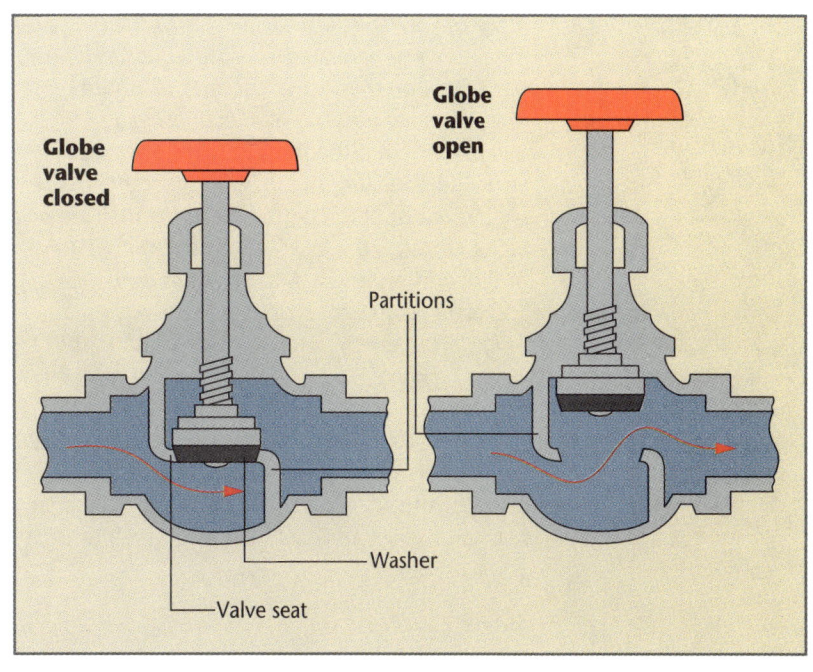

PLUMBING REPAIRS

ANGLE VALVES

An angle valve *(right)* is similar to a globe valve, except that the water inlet and outlet are at a 90° angle to each other. Water flow is less restricted than in a globe valve because the water makes only one turn instead of two. Since an angle valve eliminates the need for an elbow, these valves are often used where a pipe bends around a corner.

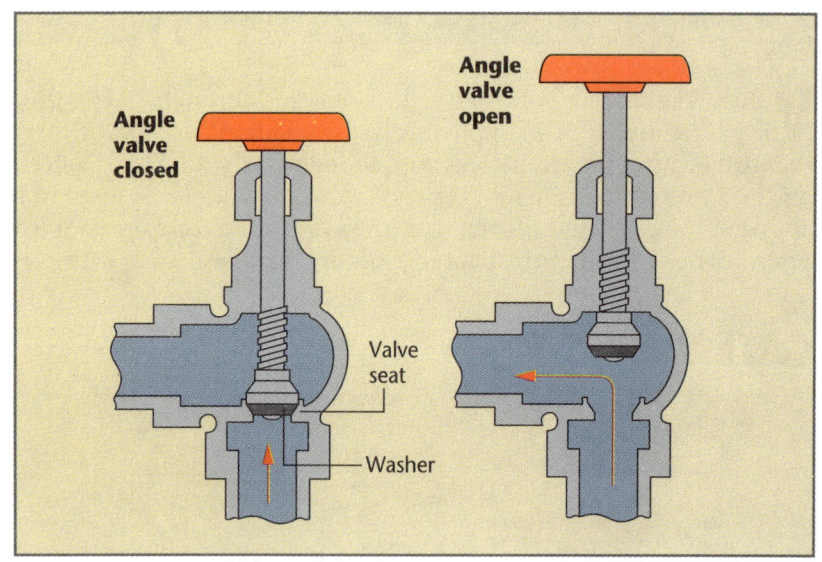

REPAIRS TO VALVES

The most common problem that occurs with valves is leaking around the stem. The usual causes are compressed packing, a faulty seat washer or disc, an obstruction in the valve, or damage to the valve seat. Despite the differing internal designs, the repairs for each valve type are basically the same. The parts of most valves are the same as those for a compression faucet, shown on page 34.

Before doing any work on a valve, shut off the water at the house shutoff valve *(page 10)*. (If the house shutoff valve itself is the one needing work, you'll need to call your local utility to turn off the water at the utility shutoff.) Open the nearest faucet to drain the pipes. Place a bucket under the valve to be worked on, to catch any remaining water.

Eliminating a leak around a valve

TOOLKIT
- Adjustable wrench
- Screwdriver
- Stiff brush

Replacing the packing and the seat washer or disc

To repair a valve, use an adjustable wrench to loosen and remove the packing nut beneath the valve handle. Examine the packing—if it's compressed, you'll need to clean away all the old packing and wrap new graphite-impregnated twine *(right)* around the base of the stem. (Other types of packing are possible, as shown on page 35.)

If that doesn't stop the leak, unscrew the valve stem and bonnet from the body. Inspect the seat washer or disc at the bottom of the stem. If it's faulty, unscrew the locknut, remove the washer or disc and replace it with an exact duplicate. You need to dress or replace a worn valve seat, *(page 36)*. Also check for an obstruction in the valve body and, if there is one, clean it away with a toothpick.

If there's still a leak, you may also need to detach the valve from its connecting pipe and clean the inside of the valve body by soaking it in vinegar or lime-dissolver and scrubbing with a stiff brush.

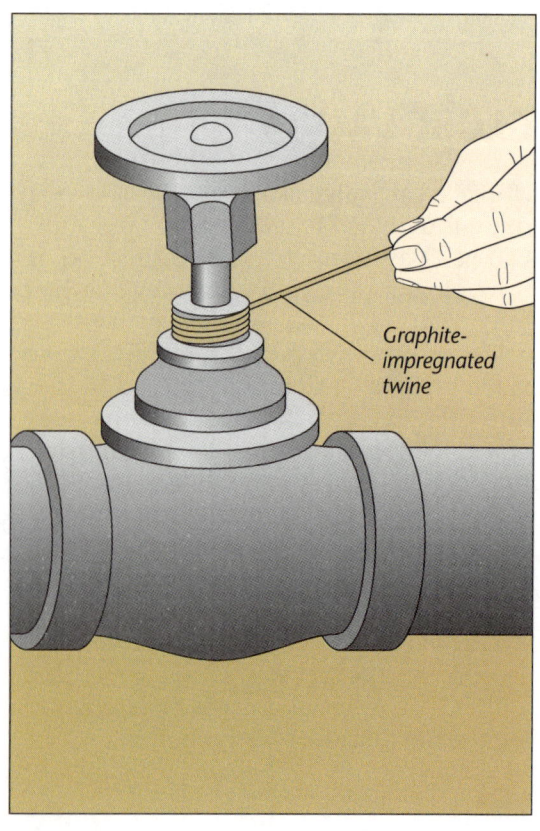

64 PLUMBING REPAIRS

WATER-PRESSURE UPS AND DOWNS

Most appliances, valves, and fixtures that use water are engineered to take 50 to 60 pounds per square inch (psi). Mains deliver water at pressures as high as 150 psi and as low as 10 psi. Too much pressure is much simpler to deal with than too little pressure. You can determine your water pressure by attaching a water-pressure gauge to an outside hose bibb or to one for a laundry hookup.

Low pressure: The symptom of low pressure is a very thin trickle of water from faucets throughout the house. Chronic low pressure is typically found in homes on hills near reservoir level. Periodic low pressure may also occur during peak service hours through no fault of the home's location. Pipes that are too small or are clogged with rust aggravate the problem by not letting through sufficient water given the low pressure. The only way of increasing water pressure to your home is to install a booster system. However, flushing the pipes (as explained below) or putting in larger sections of pipe will at least give you a greater volume of water. Consider replacing the section of pipe that leads from the outdoor utility shutoff to the house shutoff valve *(page 5)*; if it's a ¾-inch pipe, replace it with a 1-inch pipe. If you have a water meter, you can also ask your utility company to install a larger one. If you add new fixtures, you may need to install a larger main supply pipe from the point where the water enters the house to the various branches of the supply network, in order to maintain a sufficient volume of water at the fixtures farthest from the main.

High pressure: The symptoms of high pressure are loud clangs when the dishwasher shuts off or wild sprays when faucets are first turned on. High pressure usually occurs in houses on low-lying slopes of steep hills or in subdivisions where high pressure is maintained for fire protection.

If your house has particularly high water pressure, take precautions against appliance damage and floods by turning off the house shutoff valve *(page 10)* when you go on vacation, and by turning off appliance shutoff valves, especially those for a washing machine and dishwasher, when not in use.

Above-normal pressure can be cured easily and inexpensively by the installation of a pressure-reducing valve, as explained below. This valve can reduce pipe pressure from 80 psi or more down to a manageable 50 to 60 psi. If you'd like outdoor hose bibbs and sprinklers to enjoy the higher water pressure, install the pressure-reducing valve downstream from them.

Dealing with low water pressure

Flushing the pipes
A system showing early signs of clogged pipes can regain some lost pressure if the system is flushed. To do this, follow these steps: Remove and clean aerators on faucets *(page 33)*. Close the gate valve that controls the pipes you intend to clean; it may be a shutoff valve on the water heater, or the house shutoff valve *(page 10)*. Open fully the faucet at the point farthest from the valve, and open a second faucet nearer the valve. Then, plug the faucet near the valve with a rag (but don't shut it off). Reopen the gate valve and let water run full force through the farther faucet for as long as sediment continues to appear—probably only a few minutes. Finally, close the faucets, remove the rag, and replace the aerators.

Dealing with high water pressure

Installing a pressure-reducing valve
The method you'll use to install a pressure-reducing valve depends on the type of pipe—galvanized, copper, or plastic—in your plumbing system. First, assemble the valve with the pipe fittings necessary to connect the threads of the valve to the existing pipe. Then, after shutting off the water supply *(page 10)*, remove a length of pipe on the house side of the house shutoff valve long enough to accommodate the valve and the assembled fittings.

Install the valve, following the instructions for pipefitting beginning on page 11. When the work is completed, you can turn the water back on. Be sure to check for any leaks in the new connections.

To minimize the water pressure, turn the adjusting screw at the top of the valve clockwise until the pressure is low enough to end bothersome pipe noises. Be sure the valve still supplies adequate water flow to the upper floors or to far-away fixtures in the house.

PLUMBING REPAIRS

PLUMBING IMPROVEMENTS

This chapter will give you the chance to put to good use some of the skills you've learned in the previous chapters. As you plan your plumbing improvement projects, your new-found knowledge of pipefitting from the first chapter will come in handy, as will your repair abilities (from the chapter beginning on page 27). You'll learn how to replace a sink *(page 81)* or a toilet *(page 85)*, and install appliances. You'll also learn about replacing a hot water heater *(page 92)*, which may be necessary to meet growing hot water needs.

When you're shopping for any new plumbing fixtures, look for water-conserving fixtures. Buying a low-flush toilet makes sense when you realize that an ordinary toilet uses $3^{1}/_{2}$ to 7 gallons of water per flush, whereas an ultra-low flush (ULF) toilet uses only 1.6 gallons of water per flush. A family of four can cut indoor water use by up to 20% simply by replacing existing toilets with ULF toilets.

There are some plumbing improvement projects that involve roughing-in pipe. You'll find instructions on this on page 67. You'll also need to know about the venting options *(page 69)* available for your new fixtures.

Installing a fixture shutoff valve (page 76) for a sink is one of the many plumbing improvements discussed in this chapter.

ROUGHING-IN AND EXTENDING PIPE

When undertaking any plumbing improvement project, start by understanding the basics of plumbing systems *(page 5)* and pipefitting *(page 11)*. Then review the information in this section—it outlines the planning process, explores some of your options, and presents techniques and general advice for roughing-in pipes to new fixtures and water-using appliances.

THE PLANNING SEQUENCE
When plotting out any plumbing addition you must balance code restrictions, the limitations of your system's layout, design considerations, and of course, your own plumbing abilities.

Don't buy a pipe, a fitting, or a fixture until you've checked your local plumbing and building codes. Almost any improvement that adds pipe to the system will require approval from local building department officials before you start, and inspection of the work before you close the walls and floor. Learn what work you may do yourself—a few codes require that certain work be done only by licensed plumbers.

A detailed map of your present system will give you a clear picture of where it's feasible to tie into supply and drain lines, and whether the present drains and vents are adequate for the use you plan.

Starting in the basement, sketch in the main soil stack, branch drains, house drain, and accessible cleanouts; then trace the network of hot and cold supply pipes. Also, check the attic or roof for the course of the main stack and any secondary vent stacks. Determine the materials and, if possible, the diameters of all pipes. When planning your plumbing, keep in mind that while it's easy to route supply pipes to most locations, it's more difficult to put in and conceal drainage and venting pipes. To minimize cost and keep the work simple, arrange new fixtures as close to the present pipes as possible. Be sure also that both your water heater and water pressure can handle the extra load of any additions.

After reading the section on pipefitting, which begins on page 11, decide what kind of pipe you'll need—galvanized steel, copper, or plastic for the supply tubes; plastic, copper, or cast iron (in some places, galvanized steel) for drain-waste and vent (DWV) pipes.

The simplest and most cost-efficient way to add a new fixture or group of fixtures is to connect to the existing main soil stack either individually or through a branch drain. One common approach is to install a new fixture or group of fixtures above or below on the stack, piggyback style (but check codes carefully). Another plan is to place a new fixture or group of fixtures back to back with an existing group attached to the main soil stack *(below, left)*.

If your addition is planned for an area across the house from the existing plumbing, you'll probably need to run a new secondary vent stack up through the roof, and a new branch drain to the soil stack *(below, right)* or to the main house drain via an existing cleanout. Installing a new secondary vent stack and branch drain can drive up your labor and demoli-

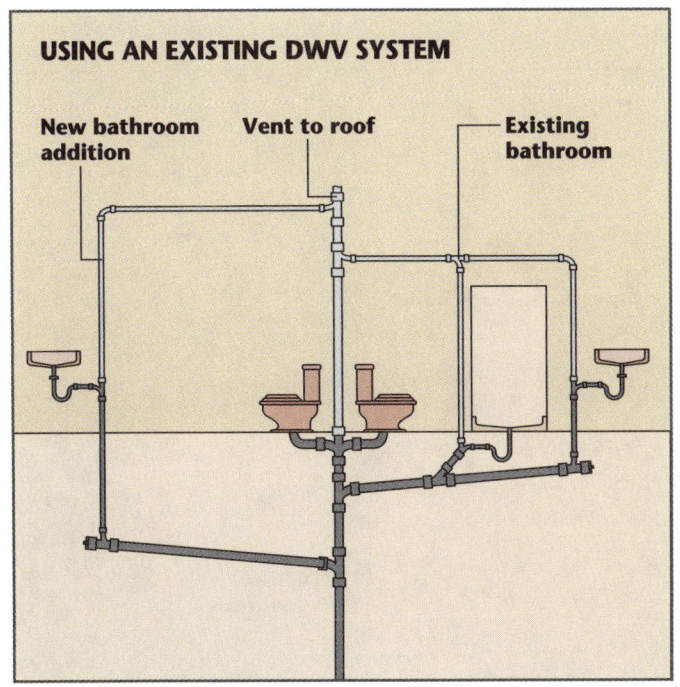

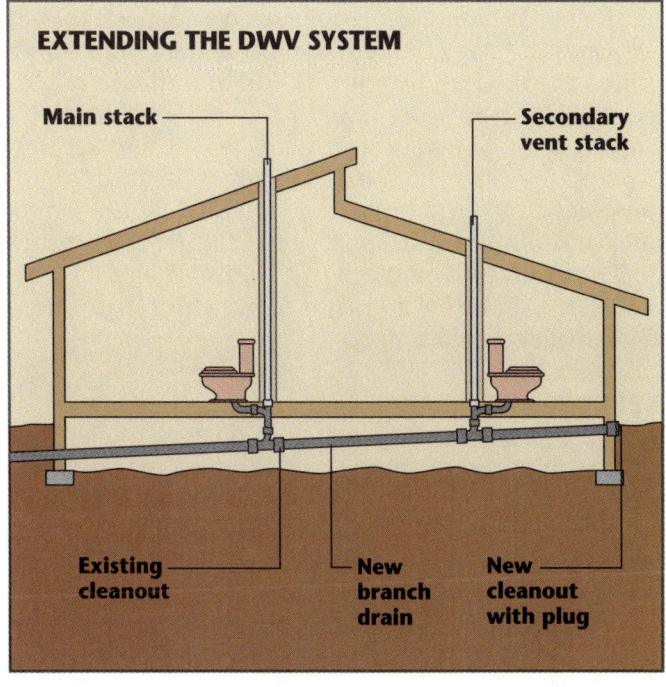

PLUMBING IMPROVEMENTS 67

tion costs. If you're remodeling, sometimes it's easier to run a new branch drain and vent stack than to cut one into the main stack that may be far from the new fixtures. Often, a bathroom sink, but not a toilet, can be tied directly into an existing branch drain.

PLUMBING CODES

Few code restrictions apply to simple extensions of hot and cold water supply pipes. Be sure your water pressure is adequate *(page 65)*. The material and diameter for supply pipes serving each new fixture or appliance should be spelled out clearly in your local code. More troublesome are the pipes that make up the DWV system; codes govern the organization of these pipes.

Three major elements of the DWV system that come under code restrictions are: **1)** the vertical main stack; **2)** horizontal branch drains; and **3)** separate vent systems. When adding a new fixture or a new bathroom, you'll need to answer these important questions:
• Is your present stack or branch drain adequate in size to tie into?
• Where can you place any new fixtures along your present DWV system?
• How will each new fixture be vented?

The plumbing code specifies minimum diameters for stacks and vents in relation to numbers of fixture units. (One fixture unit represents 7.5 gallons or 1 cubic foot of water per minute.) In the code, you'll find fixture unit ratings for all plumbing fixtures in chart form.

To determine drainpipe diameter, look up the fixture or fixtures you're considering on the code's fixture unit chart. Add up the total fixture units; then look up the drain diameter specified for that number of units.

Vent pipe sizing criteria also include length of vent and type of vent, in addition to fixture unit load. (Refer to the discussion of venting options *(opposite)*.

The maximum distance allowed between a fixture's trap and the stack or main drain that it empties into is called the critical distance. No drain outlet may be completely below the level of the trap's crown weir, shown in the illustration at right, or it would act as a siphon, draining the trap; thus, when the ideal drainpipe slope of 1/4 inch per foot is figured in, the length of that drainpipe quickly becomes limited. But if the fixture drain is vented properly within the critical distance, the drainpipe may run on indefinitely to the actual stack or main drain.

LOCATING AND EXPOSING PIPES

Before you can extend the present pipes to reach a new fixture or group of fixtures, you'll need to pinpoint where they run in walls and floors. Then, to gain elbow room, you'll need to carefully remove wall, ceiling, and floor materials in the immediate area.

By now, you should know roughly the location of the pipes you'll tie into. Here's where your system map comes in handy. Locate pipes as exactly as possible

ASK A PRO

CAN I DO IT MYSELF?

Extending supply, drain-waste, and vent pipes to a new fixture or a new fixture group requires the ability to accurately measure pipe runs, calculate DWV (drain-waste and vent) slope, and cut and join pipe and fittings. In addition, general carpentry skills and tools are needed for opening up walls or floors, and notching or drilling framing members. Improvements that involve a new soil stack or the addition of a new venting system are particularly messy and demanding.

If you have any hesitations about these tasks, consider hiring a professional to check your plans and install the drain-waste and vent system—or to rough-in all the pipes. Starting on page 11, you'll find instructions on how to join the parts of the water supply runs yourself.

from above or below. You may have to drill or cut exploratory holes to pinpoint the location of stacks, branch drains, or water supply risers inside a wall or ceiling. Once you find one riser, the other may be about 6 inches away.

To open the wall, insert a metal tape measure into the exploratory hole near the pipes you're tying into. Run the tape left until you hit a stud; note the measurement, then mark the distance on the outside of the wall covering. Then repeat the process to the right. With a carpenter's level that also shows plumb, draw vertical lines through your marks to outline the edges of the flanking studs. Then turn the level to the horizontal and connect the vertical lines above and below where you plan to tie into the pipes. A rectangle about 3 feet high should be large enough.

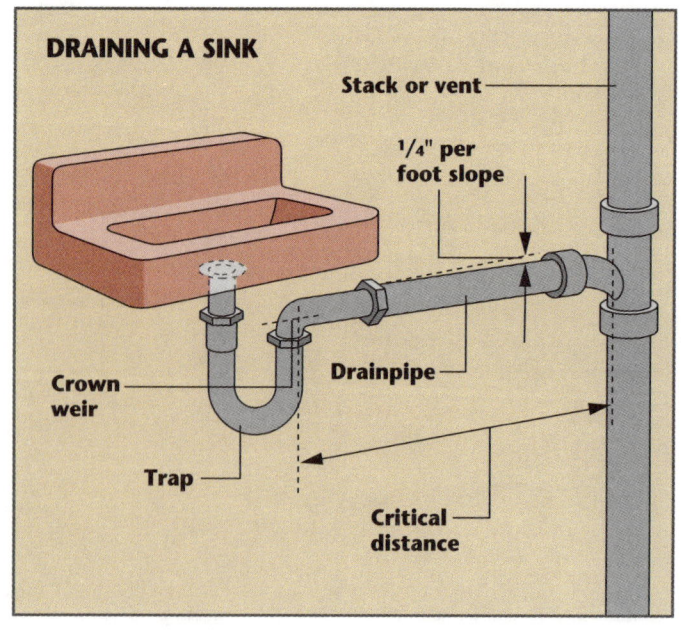

DRAINING A SINK

VENTING OPTIONS

Your four basic venting options—subject to local codes—are wet venting, back venting, individual venting, and indirect venting.

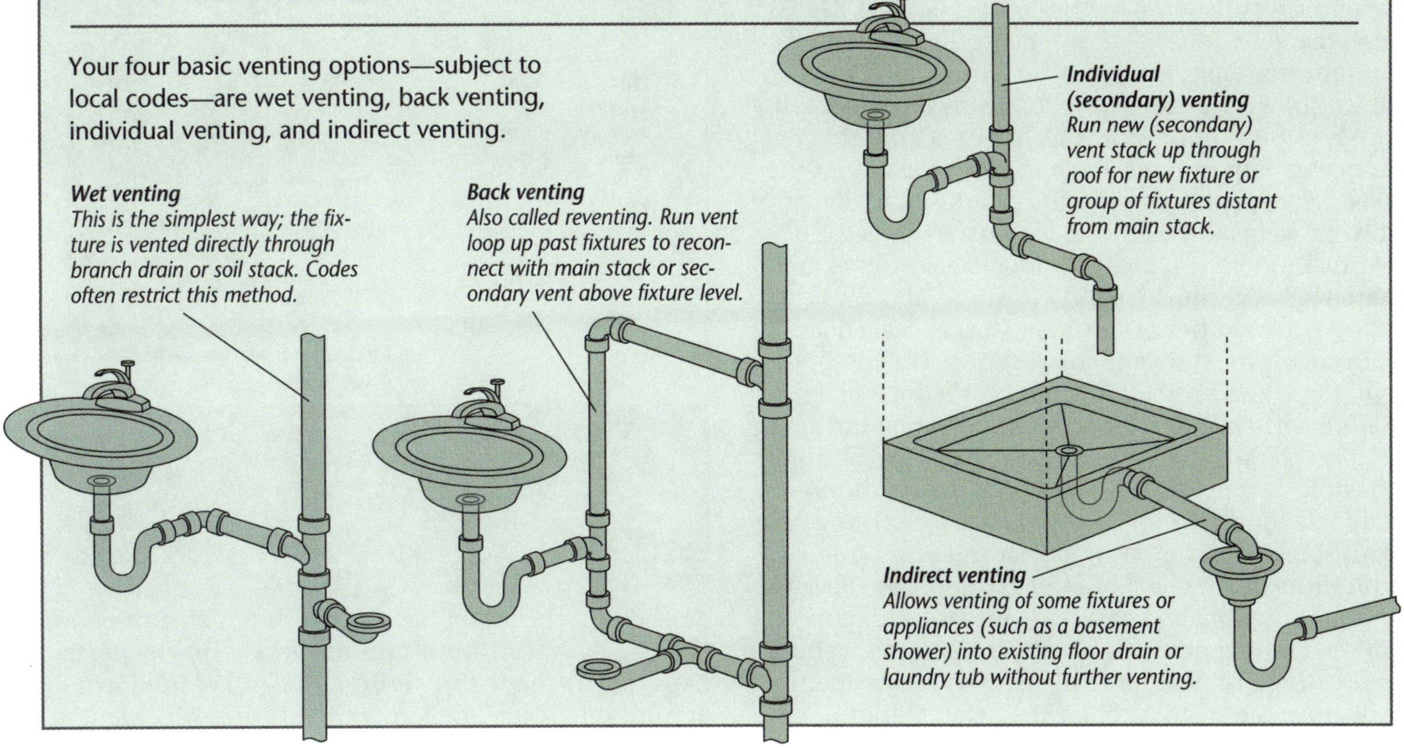

Wet venting
This is the simplest way; the fixture is vented directly through branch drain or soil stack. Codes often restrict this method.

Back venting
Also called reventing. Run vent loop up past fixtures to reconnect with main stack or secondary vent above fixture level.

Individual (secondary) venting
Run new (secondary) vent stack up through roof for new fixture or group of fixtures distant from main stack.

Indirect venting
Allows venting of some fixtures or appliances (such as a basement shower) into existing floor drain or laundry tub without further venting.

To cut into gypsum wallboard, drill small pilot holes at the four corners of your outline, then use a compass saw to cut along the lines you've marked *(below, right)*. Be careful to avoid electrical wiring.

Floors can be a messier proposition than walls—you have to tear out the floor covering (and repair it later), as well as the subfloor. If you're tying into a branch drain, try to gain access from below—from the basement for a first-floor drain, or through ceiling materials below the second floor. To open a ceiling between joists, follow the procedure for opening walls.

PIPE CONNECTIONS

Basically, tying into drain-waste, vent, and supply lines entails cutting a section out of each pipe, inserting a new sanitary fitting or supply T, and running pipes to the new fixture along preplotted lines.

Laying out the plan: Most new fixtures will include a fixture template or other roughing-in measurements telling where supply pipes and the trap exit into the drainpipe (the spot where the drainpipe enters the wall or floor) should be located on the wall or floor. Position these measurements carefully on the wall or floor where you prefer them. The combined length of the new fixture's drain and the height of its trap exit on the wall or below the floor will determine exactly where the connection to a stack or branch drain will be made.

To plot a stack connection inside a wall (for a sink), first mark the fixture's roughing-in measurements on the wall. To plot the 1/4-inch-per-foot slope for your drainpipe, run a tape measure from the center of the trap exit mark to a point at the same height on the stack. Subtract 1/4 inch per foot of this distance, and lower the mark on the stack by this amount *(page 70)*.

If you're installing a toilet, position the new toilet bend below the subflooring. Figure slope with a chalk line snapped on a parallel joist, or a string pulled taut along the proposed run.

Supply pipes are not required to slope 1/4 inch per foot in the same way that drainpipes are, but figuring in at least a slight slope allows you to drain the pipes later. Run hot and cold supply pipes parallel to each other, about 6 inches apart.

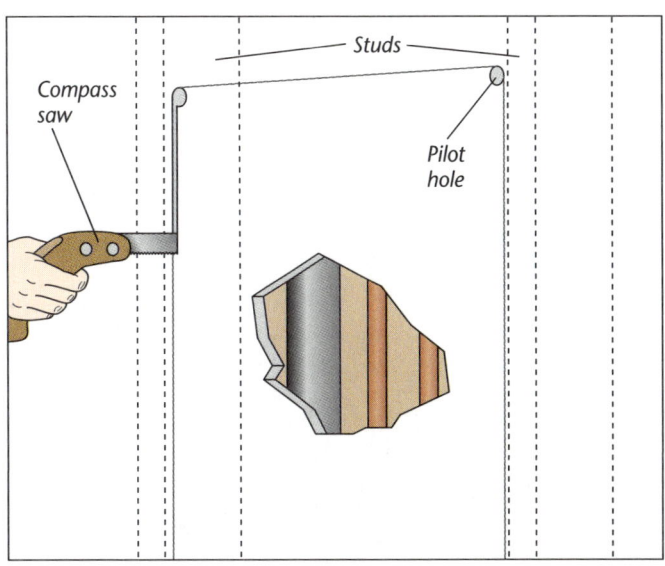

PLUMBING IMPROVEMENTS 69

If you cannot access a vent stack easily or if you wish to avoid running a new vent in a plastic DWV system, an antisiphon vent valve can be used. The 1 1/2-inch-diameter PVC valve automatically lets in air to prevent trap siphoning in wash basins, tubs, kitchen sinks, showers, and washing machine drainpipes. It closes to prevent pressures higher than atmospheric from escaping through traps. A larger version of the valve serves a toilet or an entire bathroom. The valve solvent-welds to a 1 1/2- or 3-inch-diameter vertical plastic DWV pipe 6 inches or more above the fixture trap. The valve must remain accessible, and thus can serve as an additional cleanout. Check local codes.

Drain-waste and vent connections: The method of tying into DWV pipes depends on the pipe material. Before cutting into the stack, support it by installing friction clamps *(below, left)* near the top and bottom of the hole in the wall where the pipe runs vertically.

When adding a new connection to an existing **cast-iron** stack or system, it is easiest to use no-hub cast-iron fittings. These make a tight connection. Because the slip coupling and gasket are intended to telescope on the existing pipe, it's simple to add a connection to an existing installation. Mark where the new drainpipe will enter the stack (explained above). Mark the top and bottom of where the new connection is planned, holding the fitting temporarily in position.

Wearing safety goggles, cut the stack at the marks using a reciprocating saw with a carbide blade. After cutting, file off any sharp edges. The stainless steel coupling and neoprene gasket should then be assembled on the pipe. Slide the stainless steel shield away from the sleeve gasket onto the pipe and fold back the

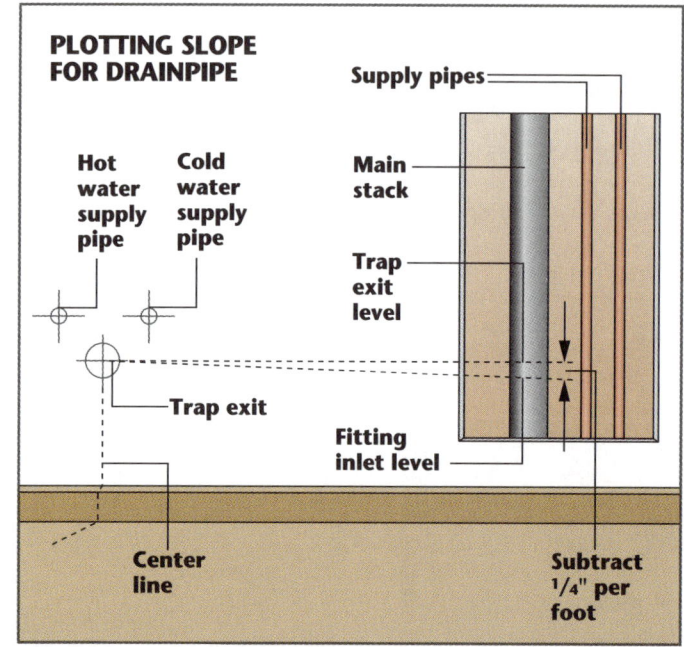

gasket's lip. The gasket has a center stop molded around the inside that will position the gasket on the pipes. Now, insert the fitting, rolling the gasket lip down around the fitting and tighten the worm-drive clamps to 60 inch-pounds of torque.

If you're dealing with **plastic** or **copper** stacks, first mark the fitting on the stack, as described previously. Then, mark the depth of the new fitting's shoulder below the top mark, and make another mark 8 inches below the fitting's bottom mark. Carefully, with a fine-toothed saw or hacksaw, cut the stack at this second set of marks.

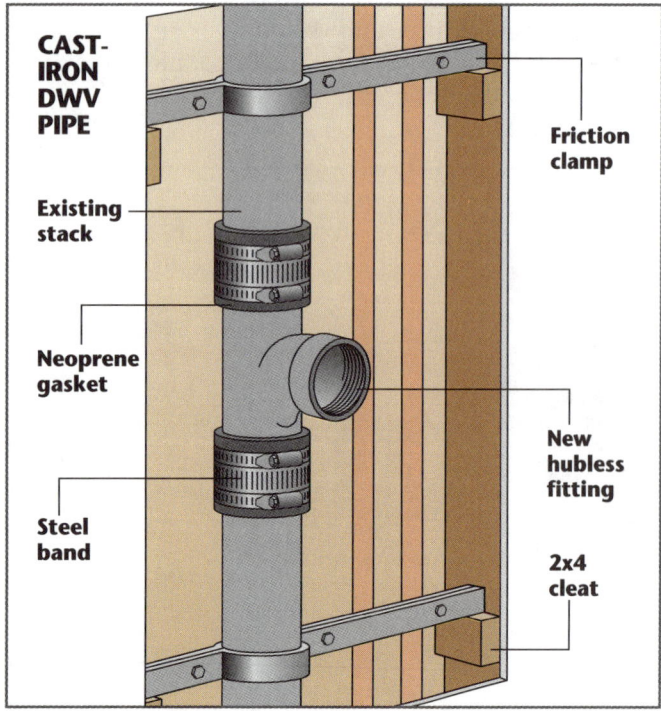

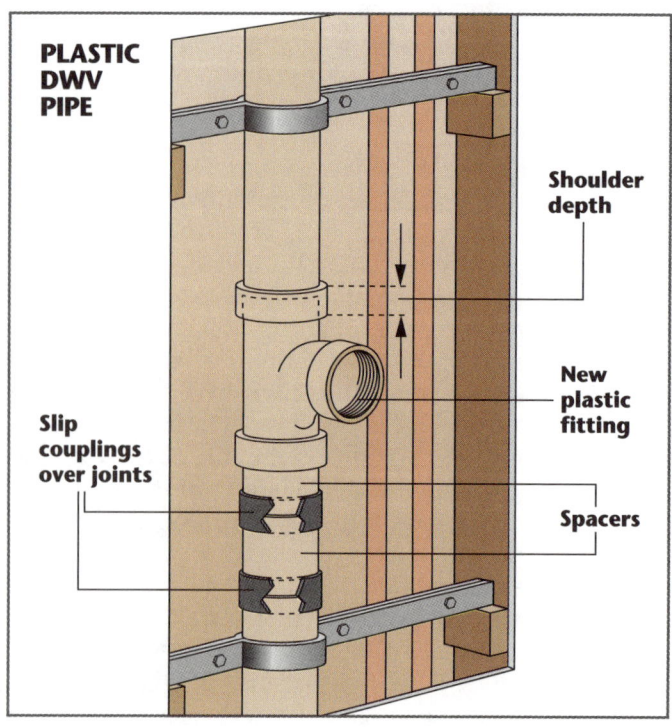

Solder or cement the new fitting to the top pipe, cut a short spacer as shown, and join it to the bottom of the fitting. Measure another spacer that fits the remaining space exactly. Slide two slip couplings onto the pipe, position the spacer, and solder or cement the two fittings over the joints *(opposite, bottom right)*. Or, in place of slip couplings on a plastic stack, use neoprene gaskets and steel bands.

Depending on the material, the same techniques outlined above are used to connect a **branch drain**, but the pipes are horizontal. Be sure the branch drain is supported adequately with pipe hangers *(page 9)* on each side of the cut.

Supply connections: Supply riser connections are made in the same way as connections to drainpipes. First, shut off the water supply at the house shutoff valve *(page 10)* and drain the pipes, if possible.

If your supply pipes are soft-temper copper or flexible plastic, simply cut the lines and insert new CPVC T-fittings *(page 12)*. Rigid plastic or hard-temper copper require that you cut out a section of pipe—about 8 inches—and, depending on the play available, add one or two spacers (nipples) and slip couplings *(below, left)*. Or, for plastic pipe, install a spacer with one threaded end and a union.

If your supply pipes are threaded galvanized steel pipes, you'll have to cut each pipe *(page 23)*, then follow it back to the nearest fitting on each end *(page 22)*. Unscrew the pipe from each fitting, using two wrenches *(page 23)*. Either install a union, the new pipe, and the T-fitting *(below, right)*, or—if you wish to change to plastic or copper at this point—add a transition fitting, a spacer, the new pipe, and a T-fitting.

NEW PIPE RUNS
With the connections made, the new DWV and supply pipes are run to the new fixture location, as marked earlier. If you wish to change pipe type—say from cast iron or copper to plastic—it's a matter of inserting the appropriate adapter at the fitting end. (Transition fittings are described under the various types of pipe, starting on page 11.)

Ideally, pipes should always run parallel to framing members and between them. At some point you'll probably have to use one of the following techniques to go through studs or joists. (NOTE: Before cutting any joists or studs, check your local building code.)

If a pipe run hits a joist near its center, you can normally drill a hole, as long as the diameter is no greater than one-third the depth of the joist. If a pipe run hits a joist near its top or bottom, a notch may accommodate it. The depth of the notch must be no greater than 1/6 the depth of the joist, and the notch cannot be located in the middle third of the span. Top-notched joists should have lengths of 2x2 wooden cleats nailed in place under the notch on both sides of the joist to give added support. Joists notched at the bottom should have either a steel strap or a 2x2 cleat nailed on. (Refer to the illustration on page 72 for the different types of pipe-joist junctions.)

You might sometimes need to cut an entire section out of a joist to accommodate a DWV section. Reinforce that section by using doubled headers on both sides of the cut *(page 72, middle)*.

Refer to the illustration on the next page for the best way to pass pipe through wall studs. You can drill a hole up to 40% of the stud depth in bearing walls (those that

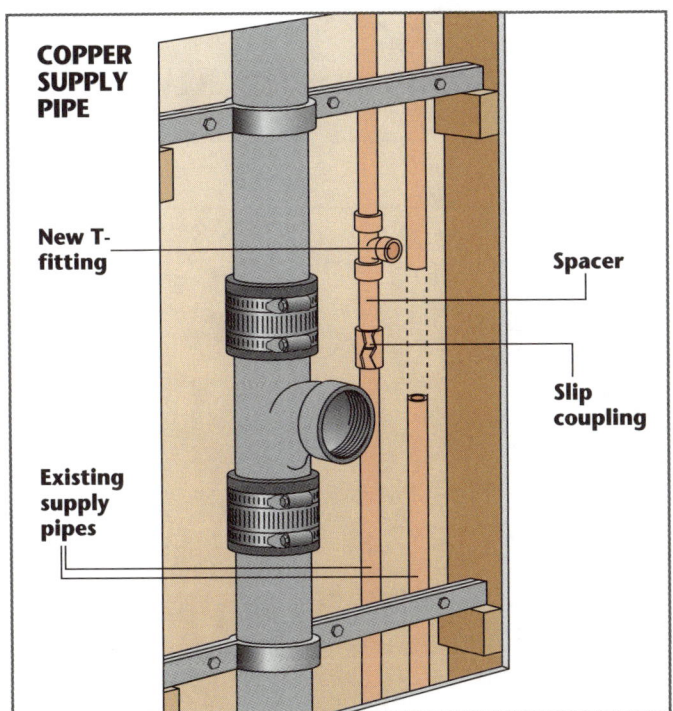

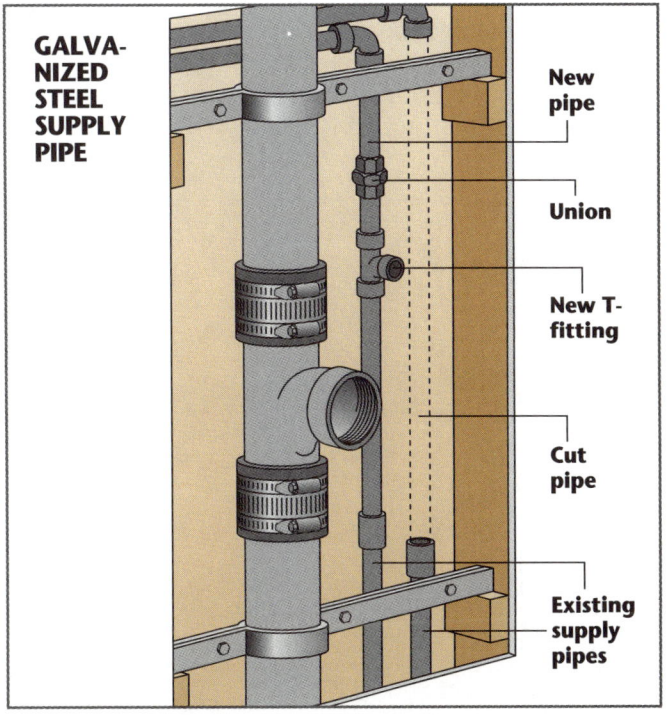

PLUMBING IMPROVEMENTS 71

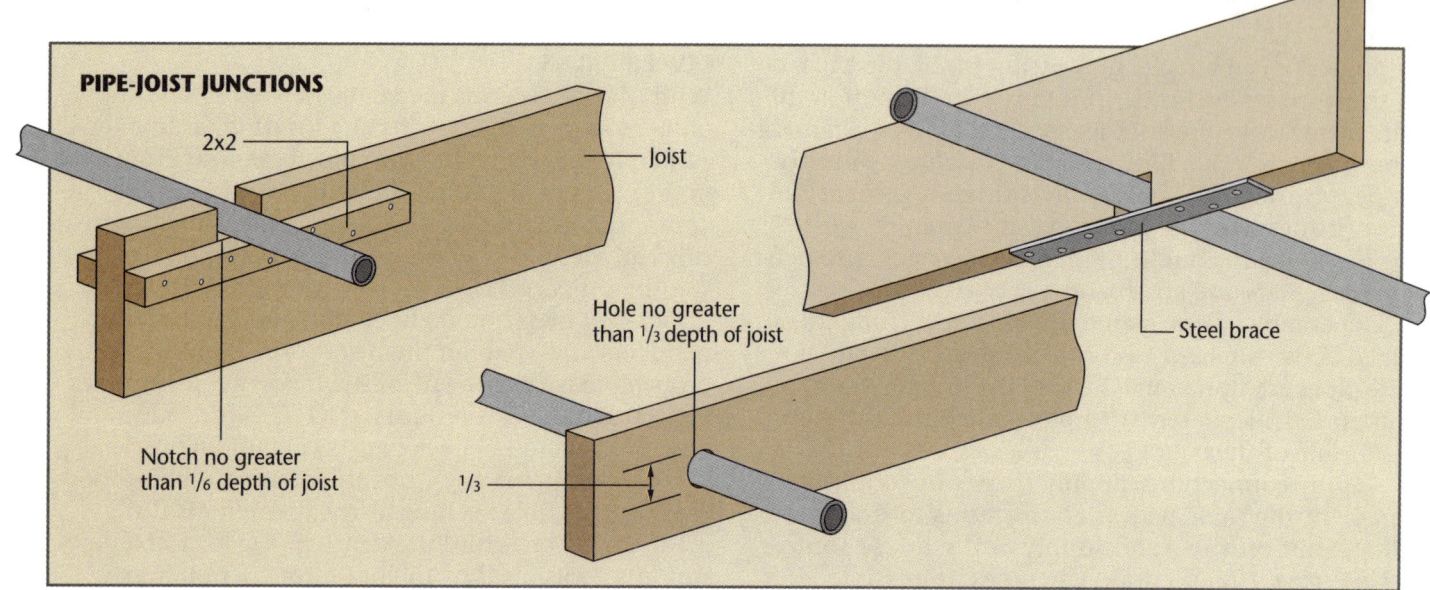

support joists or rafters above), and up to 60% in non-bearing walls. You can also drill up to 60% in a bearing wall if you nail another stud to the first stud for strength. Holes should be centered. You can notch up to 25% of the stud depth in bearing walls and up to 40% in non-bearing walls. Notches should be braced with steel ties.

Running pipes outside the wall requires no notching and leaves less wall that needs to be patched. Use blocks of wood to temporarily support pipes; then fasten pipe straps to wall studs *(opposite page, top right)*. You can hide the pipes by building cabinets, a closet, a vanity, or shelves over them.

To thicken a wall, new studs and wall materials may be erected in front of the old wall, both to cover pipes hung on the wall and to accommodate oversize DWV fittings, such as a new stack. Build out the entire wall, or thicken the lower portion only, leaving a storage ledge or shelf above the pipes.

Building up the floor will cover a new branch drain. Either thicken the floor with furring strips around the

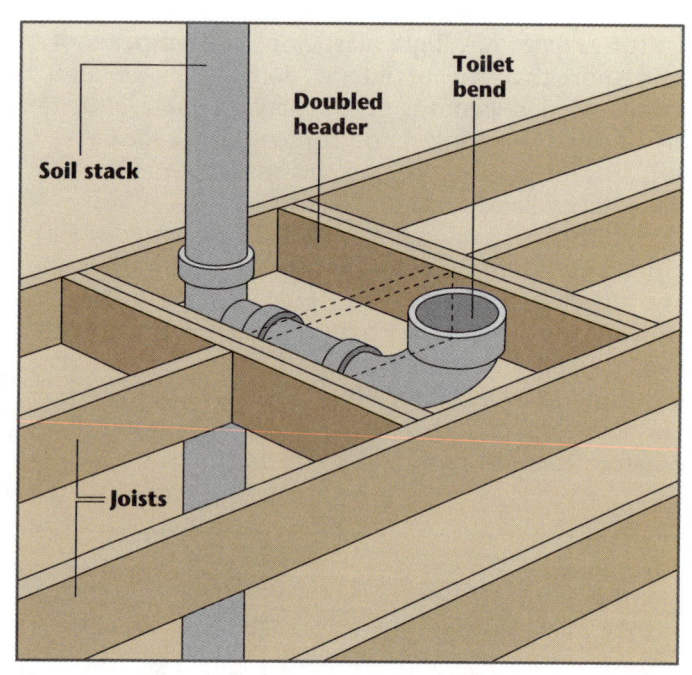

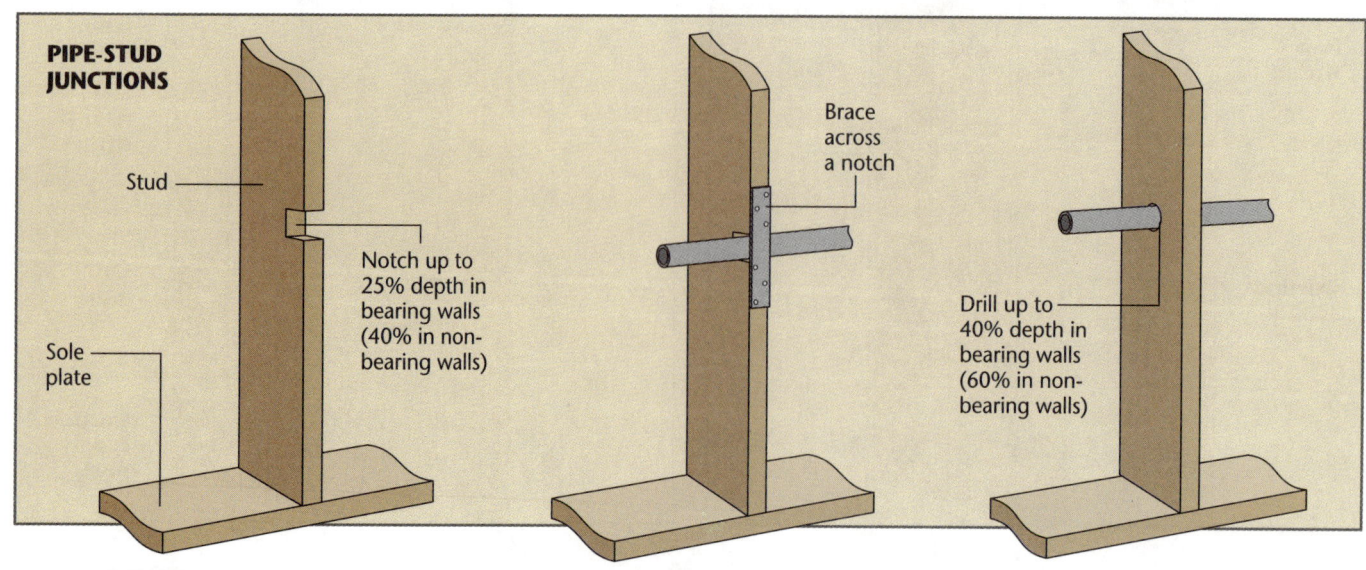

new pipes, or build a platform over the pipes for the fixture or appliance.

Leave your new pipe fittings and runs uncovered for a few days to check for leaks. Then patch the wall, ceiling, or floor. Just remember that this raised floor will affect doors and transitions to other rooms' floors.

ROUGHING-IN FIXTURES

Starting below, you'll find installation notes for roughing-in new fixtures that require tying into your present DWV and supply systems, or extending them.

A sink is the easiest fixture to install, whereas a toilet is a bit more complicated. Instructions for installing a sink are given below; those for a toilet are found on page 74. Because tub or shower installation is difficult from both the framing and plumbing perspectives, this will not be covered here.

Roughing-in a bathroom sink: As shown below, a bathroom sink is fairly easy to install and has little effect on a drain's efficiency. Common methods are back-to-back (requires little pipe) and within a vanity cabinet (hides pipe runs). A sink can often be wet-vented if it's within the critical distance; otherwise it should be back-vented. Or, if codes permit, use an automatic antisiphon vent valve *(page 70)* in place of a through-the-roof vent. Pipes you will require include: 1/2" hot and cold supply stubouts; shutoff valves; transition fittings, if necessary; and flexible riser tubes above shutoff valves. Water hammer arresters should be installed on the hot and cold supply pipes. While a long-sweep elbow can be used, a Y-fitting with a cleanout plug and a 45° elbow is better.

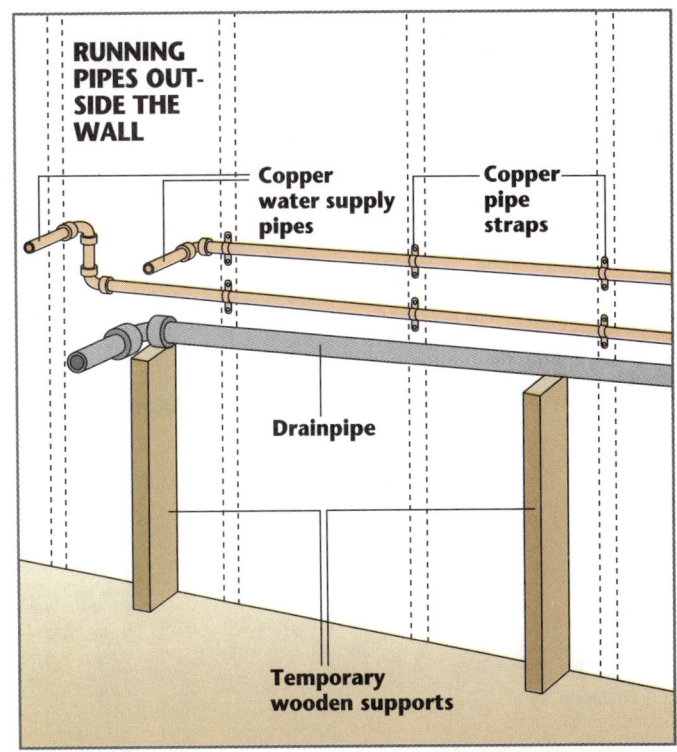

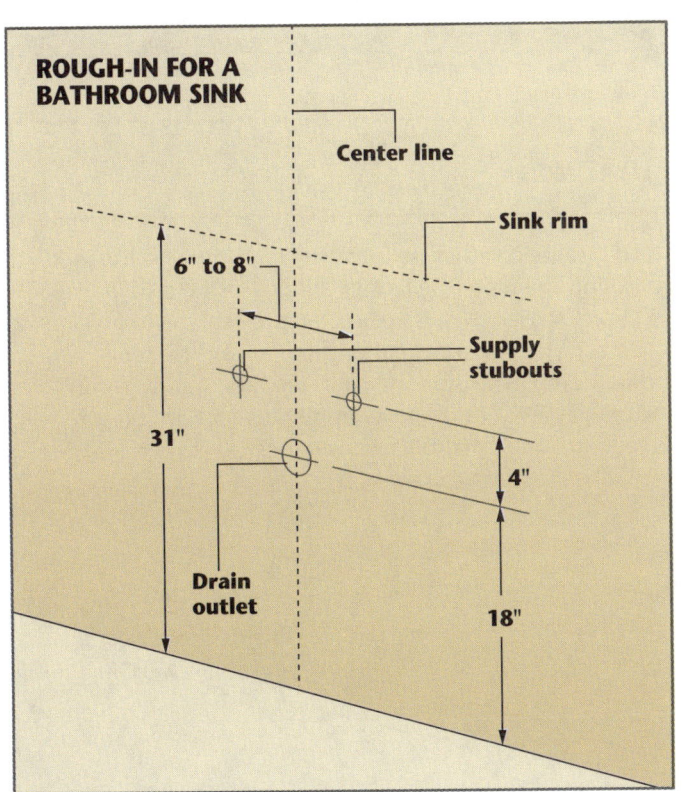

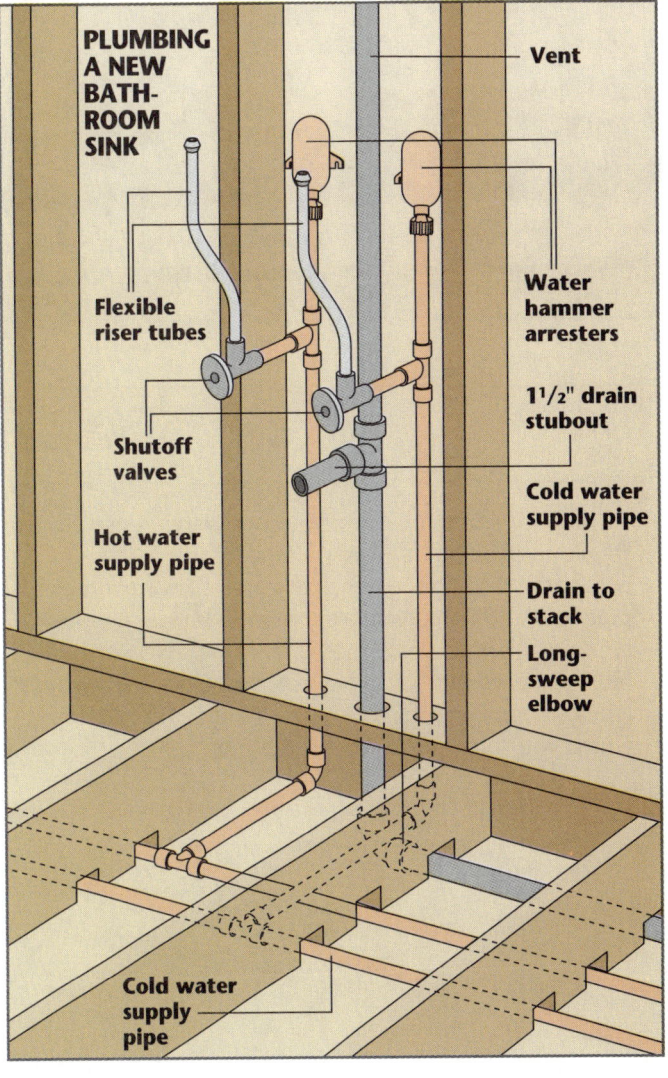

PLUMBING IMPROVEMENTS 73

Roughing-in a toilet: The toilet is the single most troublesome fixture to install in a house, because it requires its own vent (2 inches minimum) and at least a 3-inch drain. If it's on a branch drain, a toilet can't be upstream from a sink or shower.

The toilet bend and toilet floor flange must be roughed-in first, as shown in the illustration below, left; the floor flange must be then positioned at the level of the eventual finished floor (otherwise you could be faced with a toilet that leaks at its base), as shown below, right. Pipes that are required for roughing-in a toilet include: 1/2-inch riser tube; a cold water supply stubout with shutoff valve; and a flexible riser tube above the shutoff valve.

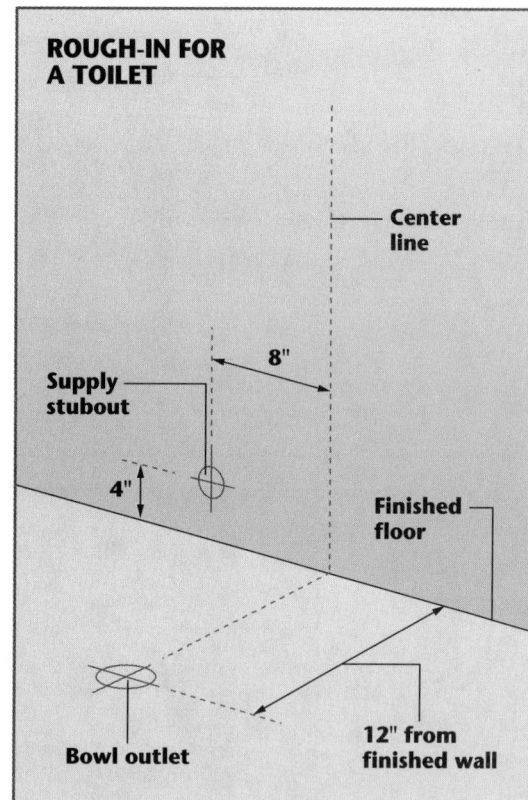

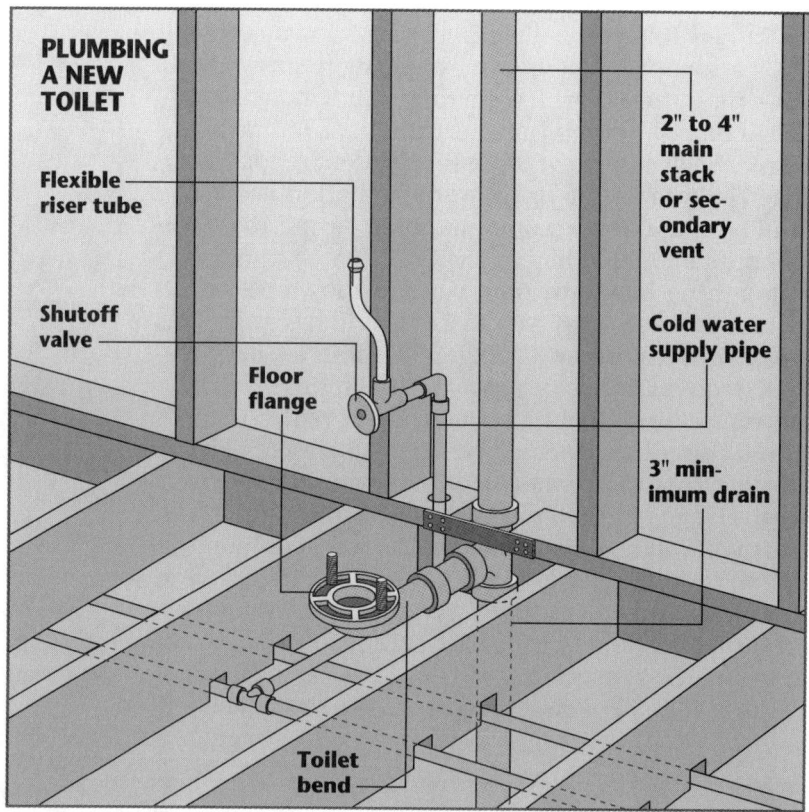

KEEPING THE HEAT IN

To minimize heat loss from your hot water system, you should insulate all hot water pipes—especially those that pass through unheated or drafty areas. Several types of pipe insulation are available; two of the most common are shown below. Polyethylene foam jackets fit around most standard pipes and are fastened with tape. Another type of insulation is foil-backed, self-adhesive foam tape, which you spiral-wrap around the pipe.

To install polyethylene foam jackets, simply fit the jacket around the pipe (the jackets are pre-slit); cut to fit at T-junctions and elbows. Use duct tape to seal the joints.

Before you apply insulation tape, clean the pipe with a mild detergent solution. Wrap the tape snugly around the pipe at a 45° angle, overlapping as you go. Be sure to redo any tape that buckles or gapes.

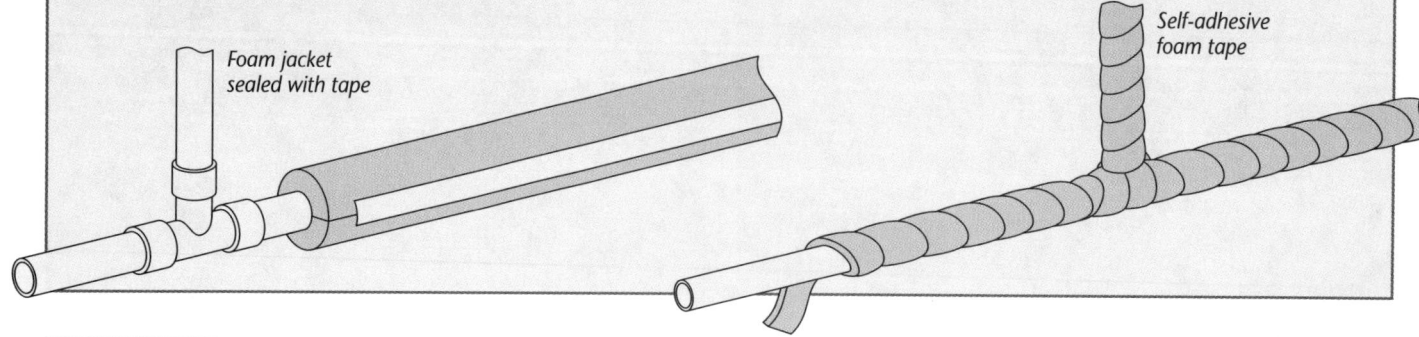

ADDING SHUTOFF VALVES

If your house is not outfitted with shutoff valves, or if you have a shutoff valve that no longer works efficiently, you'll find that installing a new one is a not-too-difficult task and one that will make future repairs much easier.

A shutoff valve simplifies turning off the water supply to a fixture for repairs or in case of an emergency. You just turn the valve handle clockwise until it's fully closed, open the faucet(s) to drain the pipes, then go ahead with your job.

Make a quick survey of your home to determine your shutoff valve needs. Every sink, tub, shower, and washing machine should have shutoffs for both hot and cold water pipes. A toilet and a water heater require only one shutoff valve (because they use only cold water), and a dishwasher needs only one shutoff valve because it uses only hot water. The shutoff for the dishwasher is usually located in the compartment of an adjoining sink. The other shutoffs are located right at the fixture or appliance.

When you shop for a shutoff valve, you'll need to choose either an angled valve or a straight one. Angled valves are used when the supply pipe, called a stubout, comes from the wall; straight valves are used for pipes that come up from the floor. For tubs and showers, choose a globe valve instead of a gate valve *(page 63)*—it's more reliable, it's easier to repair, and, unlike a gate valve, it can control the amount of water flow.

Select a shutoff valve that fits the existing pipe and is compatible with the pipe material. Copper tubing takes brass valves; iron and plastic pipe use iron and plastic valves, respectively. You can use transition fittings if the pipes aren't compatible; these fittings allow you to change materials of the valve and its connecting pipe (from galvanized steel to plastic, for example). You may want a chrome-finished valve if it will be visible.

The kind of stubout you have dictates the kind of fittings you need. A threaded pipe naturally requires a threaded valve; a copper stubout takes a fitting that's soldered *(page 20)* at one end and threaded at the other. A CPVC shutoff valve has an O-ring fitting that slips over the outside of any tube-sized stubout and it easily hand-tightens to connect a flexible riser tube to a fixture's hot or cold water supply. This type of valve fits copper tubing, CPVC, and PB. A push-in, hand-tightened connection accepts $3/8$-inch outside diameter polybutylene riser tubes leading to the fixture. Made in straight and angled versions, it's ideal for connecting washbasins, kitchen sinks, laundry tubs, and toilets.

Lengths of flexible riser tubing *(above)*, sometimes called flexible connectors, save you the trouble of piecing pipe together to join the valve to the fixture. These come in plain copper, chrome-plated copper, or plastic

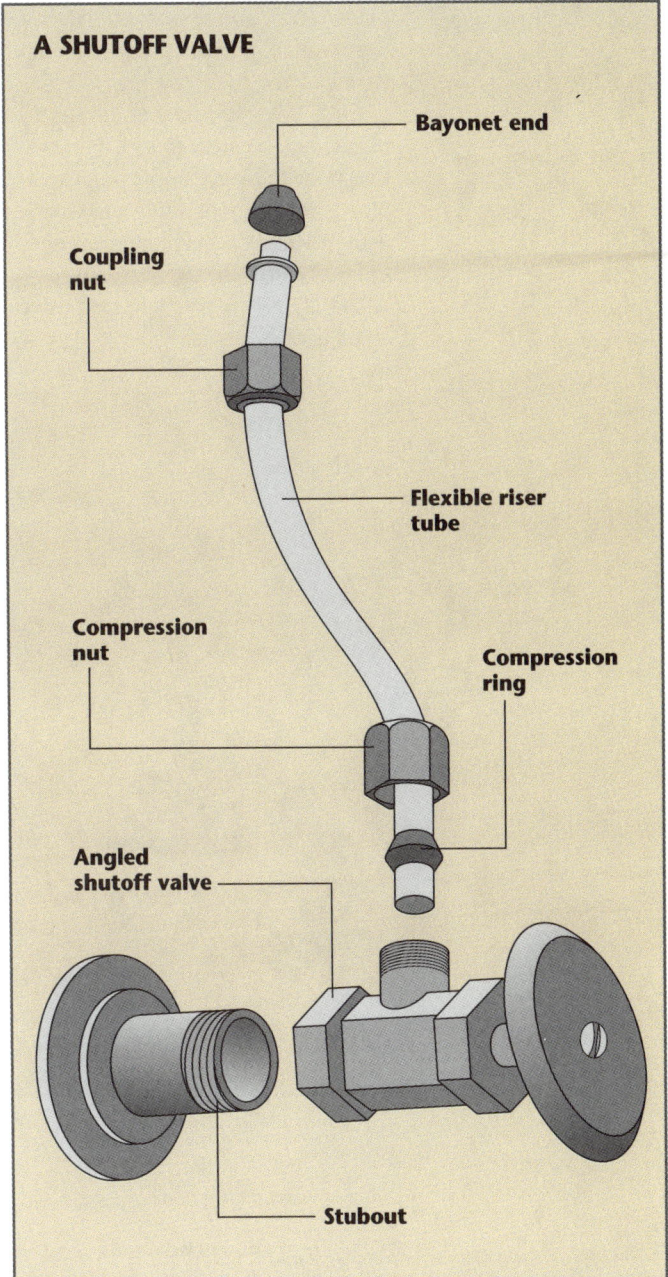

A SHUTOFF VALVE

and are available in a variety of lengths, so measure the size you need.

Flexible riser tubing for a sink or basin has a conical rubber gasket on one end and is $1/2$ inch in diameter; tubing for a toilet has a large, flat gasket on one end and a diameter of $7/16$ inch.

Unless you buy a kit that contains a shutoff valve and tubing already joined, you join them yourself with a built-in compression fitting.

PLUMBING IMPROVEMENTS

Installing a shutoff valve

TOOLKIT
- Pipe cutter or fine-toothed hacksaw
- Adjustable wrench
- Pipe wrenches or butane/propane torch
- Basin wrench

1. Removing the supply pipe
CAUTION: Before doing any work, turn off the water at the house shutoff valve *(page 10)*; open a faucet below where you're working to relieve the pressure, and put a bucket under the pipes you're removing.

Cut a 1/2" section out of the supply pipe near the elbow *(right)*. If you're dealing with plastic or copper, use a pipe cutter, if there is room to work; with galvanized steel pipe, you'll need a fine-toothed hacksaw. To free the pipe, detach the shank coupling nuts using an adjustable wrench.

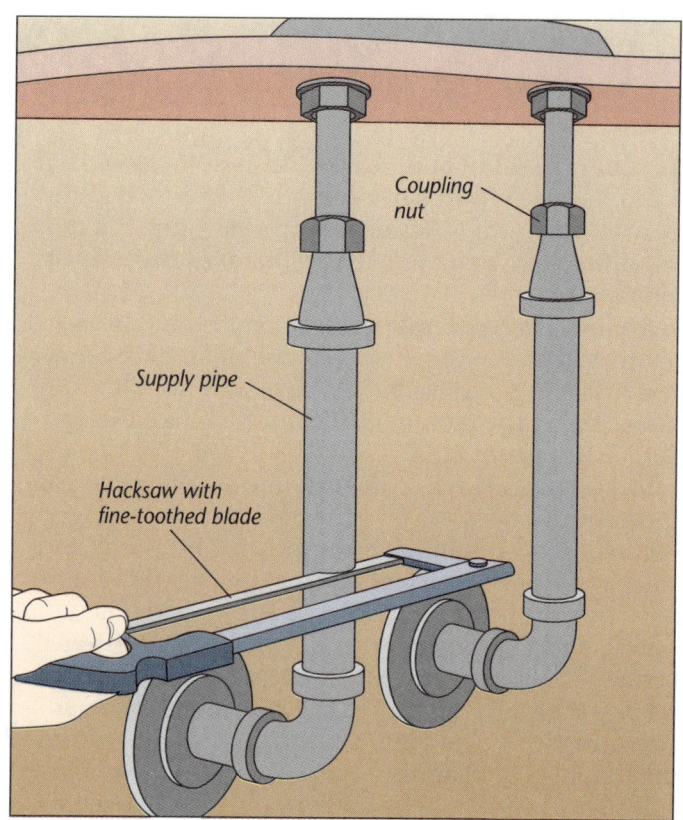

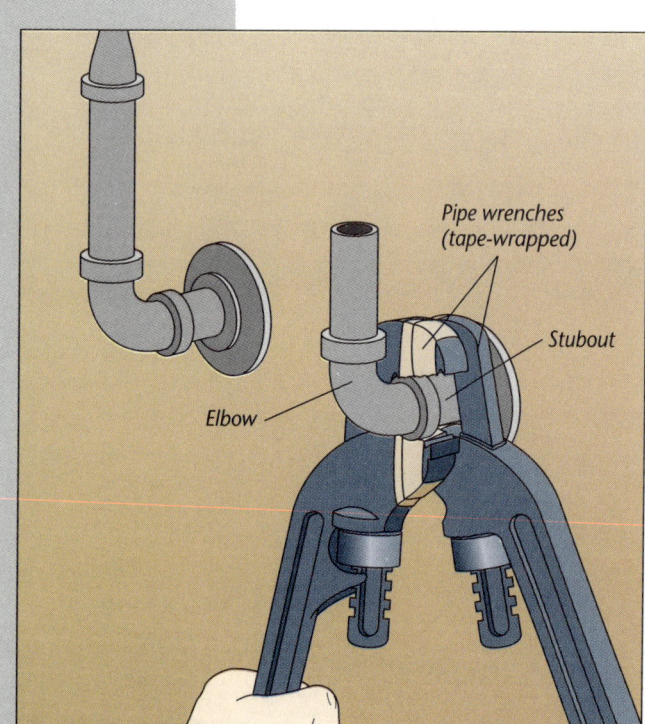

2. Detaching the elbow from the stubout
For galvanized steel pipe, use two tape-wrapped pipe wrenches *(left)* to loosen the connection. If working with copper tubing, unfasten any mechanical connections or use a torch to melt soldered joints *(page 18)*. Plastic pipe can be cut with a hacksaw. Remove everything attached to the stubout.

3. Installing the shutoff valve
Clean and prepare the exposed end of the stubout to accept the appropriate fitting. (If the stubout isn't threaded, solder or cement a fitting to it.) Screw the shutoff valve to the fitting or pipe after applying pipe-joint compound to the threads. Hand-tighten the connection, then use tape-wrapped wrenches to snug it up. Try to line up the valve outlet directly below the faucet inlet.

Cut and bend the flexible riser tube to fit its head into the fixture inlet, and its bottom end inside the shutoff valve outlet *(right)*. Fasten the coupling nut to the faucet inlet shank with a basin wrench. Secure the compression nut to the valve with an adjustable wrench. Finally, turn on the water and check for leaks.

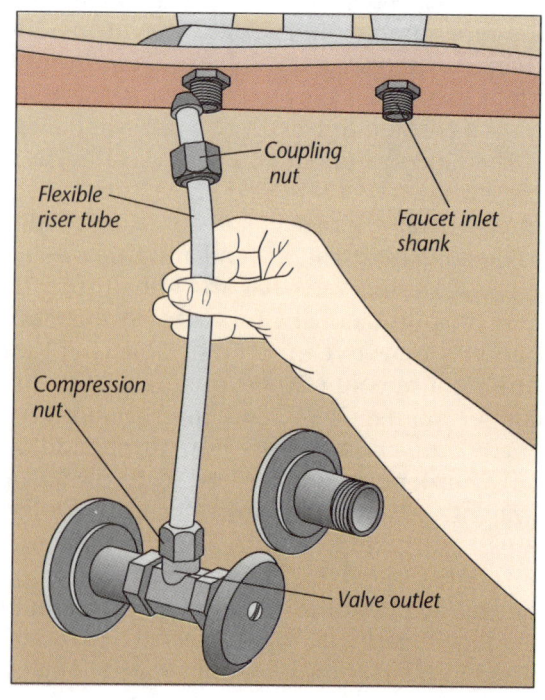

PLUMBING IMPROVEMENTS

INSTALLING FAUCETS AND SHOWER HEADS

When shopping for a new deck-mounted faucet, you'll find the selection staggering. You can choose from an assortment of single-handle, washerless faucets—disc, ball, or cartridge—and from antique-style reproductions to futuristic compression faucets. Even with all these styles, our concise directions will help make the actual installation a relatively simple process.

If you want to replace a bathtub faucet or shower head, turn to page 79 for some useful guidelines.

Adding a faucet or hose bibb on the exterior wall of your house is a fairly straightforward procedure that will pay off in convenience when you need a handy source of water outside your house. Turn to page 80 for detailed instructions.

DECK-MOUNTED SINK FAUCETS

All models are interchangeable as long as the faucet's inlet shanks are spaced to fit the holes of the sink you'll be mounting it on. If possible, take the old faucet along with you when you buy a replacement. Also, measure the diameter of the supply pipes. Choose a new unit that comes with clear installation instructions, and a well-known brand that will have repair kits and replacement parts available for future use.

Some faucets come with the copper or plastic flexible tubing for the water supply already attached. Plain copper flexible tubing is most often used where it will be concealed by a cabinet. If the tubing will be in plain view, or the existing tubes are damaged or scratched, you might want to buy new chrome-plated copper tubes, or plastic ones, along with the faucet.

Because the lengths of tubing are flexible, you can bend them to get from the faucet inlet shanks to the shutoff valves on the stubouts at the wall. If necessary, use a hacksaw or small pipe cutter to cut the tubing to the correct length. If the faucet doesn't come with flexible tubing attached, fasten the new or existing flexible tubing first to the shutoff valves under the sink and then to the faucet inlet shanks. For more information about flexible tubing, see page 17.

Replacing a deck-mounted faucet

TOOLKIT
- Basin wrench
- Adjustable wrench (optional)

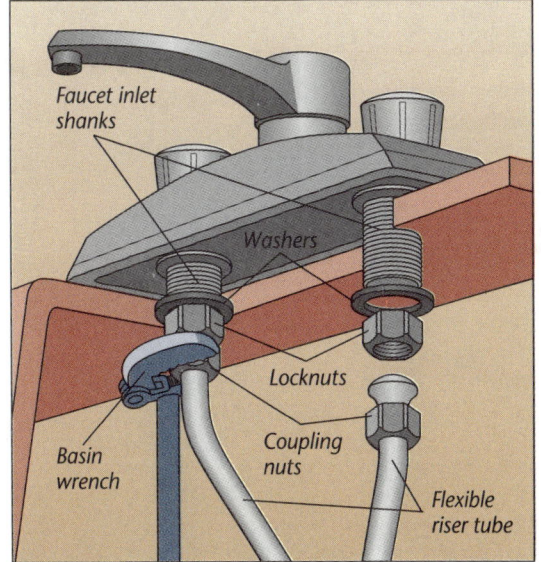

1. Removing the old faucet
CAUTION: Before doing any work on a faucet, turn off the water at the fixture shutoff valves or the house shutoff valve (page 10). Open a faucet lower down to drain the pipes; have a bucket ready.

Space is limited under the sink, so use a basin wrench to remove the nuts connecting the flexible riser tube to the faucet inlet shanks (above). Then, loosen the locknuts on both shanks and remove the locknuts and the washers. Lift out the faucet.

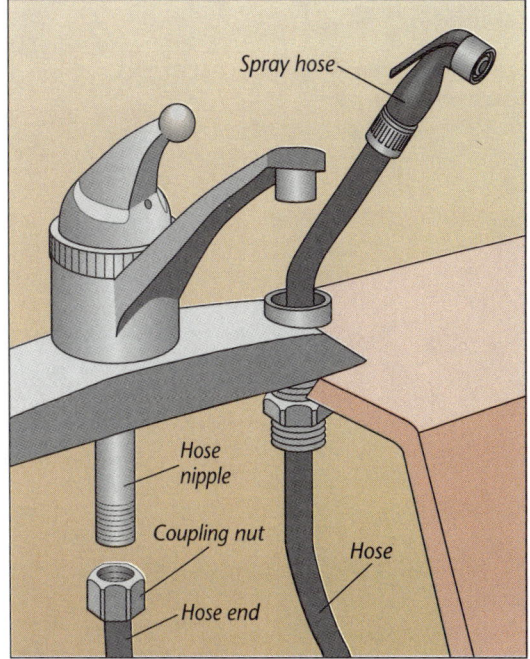

2. Removing other parts of the fixture
If you're working on a bathroom sink that has a pop-up assembly, remove the stopper (page 46).

On a kitchen sink with a spray hose attachment, use an adjustable or basin wrench to undo the nut connecting the hose to the hose nipple under the faucet body (above).

PLUMBING IMPROVEMENTS

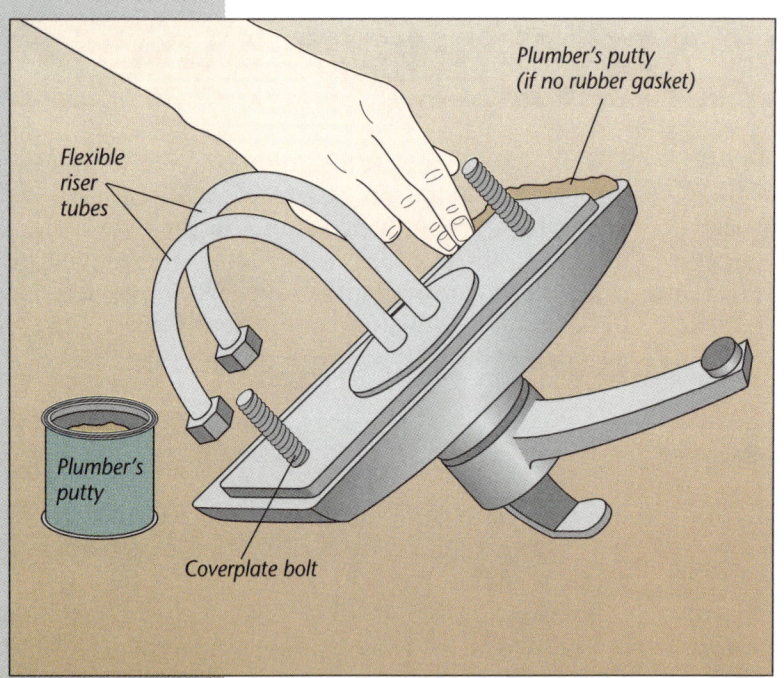
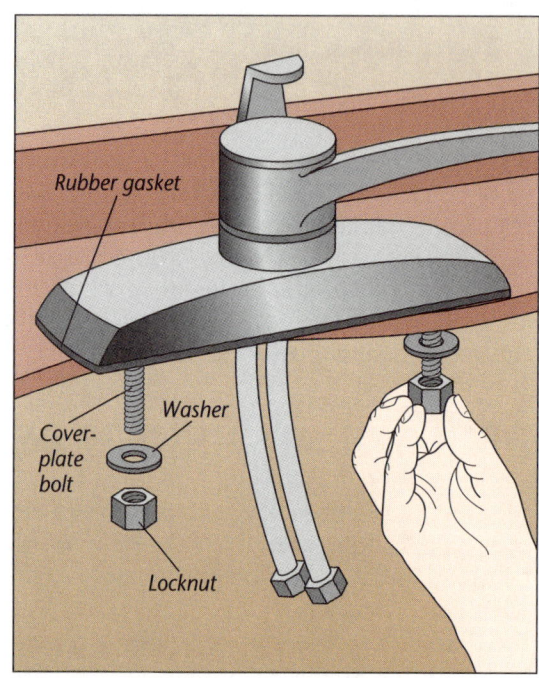

3 **Installing a new faucet**
Clean the surface where the new faucet will sit. Most faucets have a rubber gasket on the bottom; if yours doesn't have one, apply plumber's putty before you set the faucet in position.

If your faucet comes with pre-attached plastic or copper flexible riser tubes *(above, left)*, carefully straighten the riser tubes before you mount the faucet; feed the tubing through the middle sink hole. Press the faucet onto the sink and securely bolt it in place *(above, right)*. If your sink has a spray hose, attach it next *(page 43)*.

For a faucet like the one shown in step 1, screw the washers and locknuts onto the faucet inlet shanks by hand; tighten further with a basin wrench.

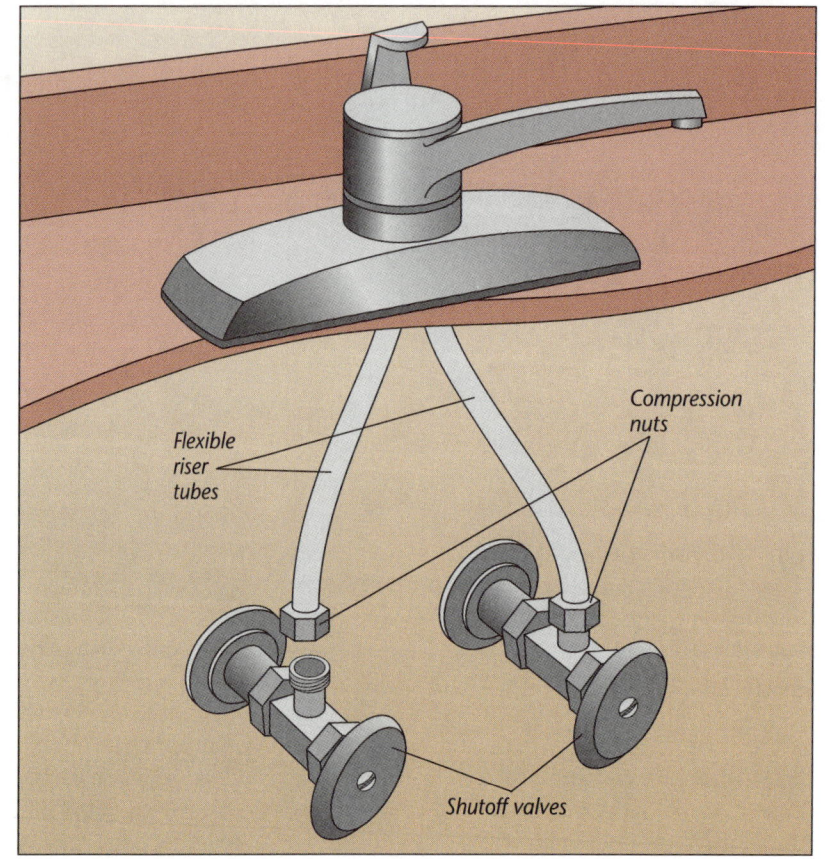

4 **Connecting the riser tubes**
Connect the flexible riser tubes *(left)*, gently bending them to meet the shutoff valves. (If there are no shutoffs, install them as shown on page 76.) Join the riser tubes to the shutoff valves, using compression nuts or flared fittings *(page 21)*. Turn on the water and check for leaks.

TUB FAUCETS, DIVERTERS AND SHOWER HEADS

Like sink faucets, tub faucets can be either compression-style or washerless—both types are illustrated below. Either way, water is directed from the faucet to the tub or shower head by a diverter valve; some, as shown below, have built-in diverter valves, while others have a knob on the tub spout. This latter type is easy to replace: Grip the old spout with a tape-wrapped pipe wrench and turn counterclockwise. Hand-tighten the new spout into place.

Like a tub spout, a shower head simply hand-screws onto the shower arm stubout. Before installing a new tub spout or shower head, clean the pipe threads and apply pipe-joint compound to prevent leaks. Refer to page 44 for guidelines on how to fix a leaking shower.

Water-saving shower heads can also be installed. Low-flow shower heads and flow restrictors can reduce water to under 2 1/2 gallons per minute, compared with the 5 to 8 gallons per minute flow of standard fixtures.

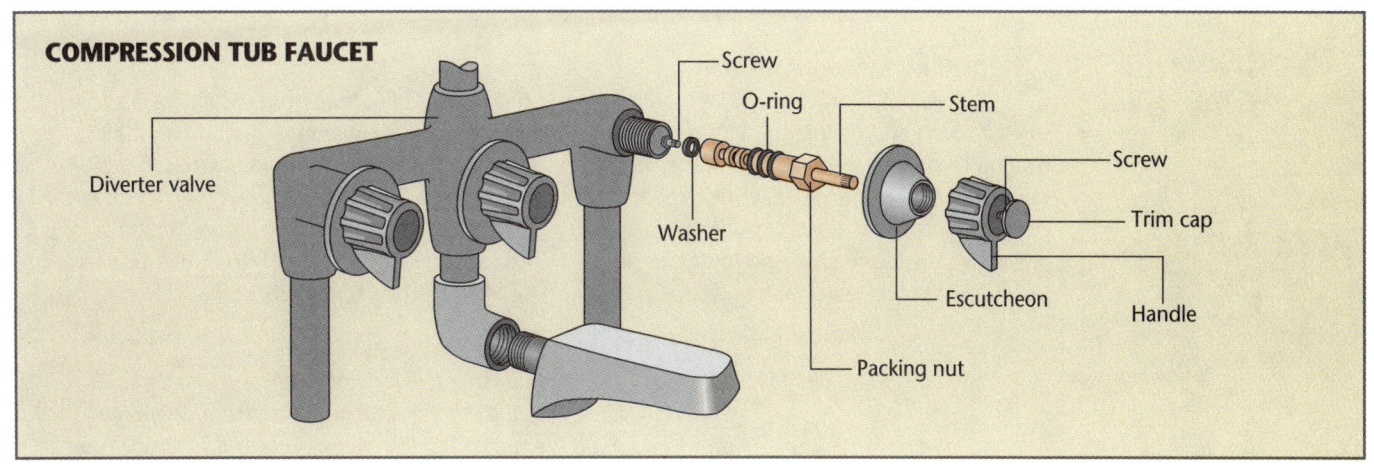

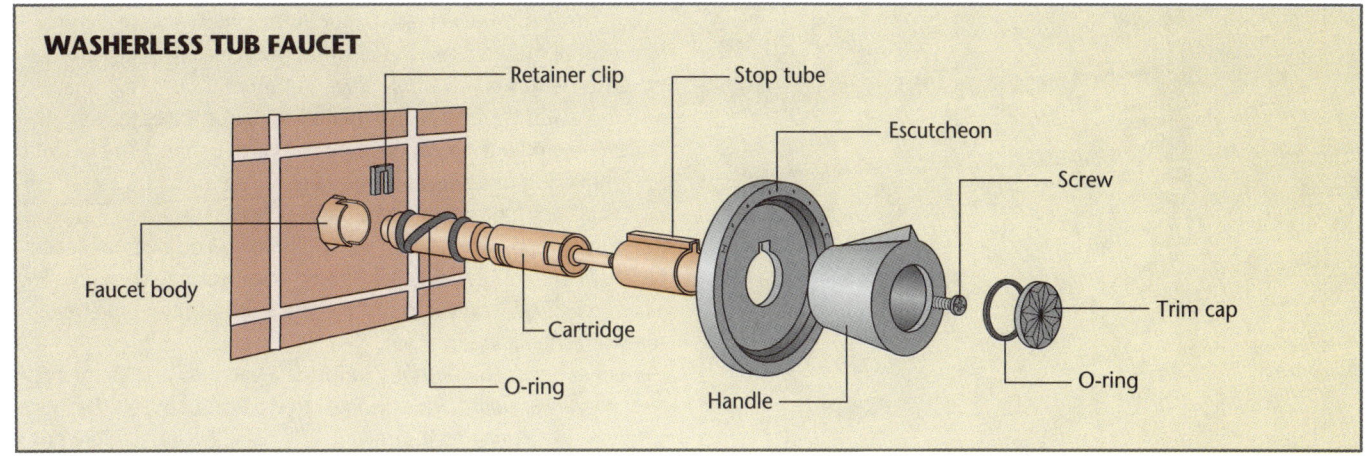

Replacing a tub faucet

TOOLKIT
- Ball-peen hammer and cold chisel
- Socket wrench

Removing the old faucet and installing a new one

CAUTION: Before you begin any work, turn off the water at the house shutoff valve *(page 10)*. Open a faucet lower down to drain the pipes.

To disassemble the faucet, refer to the section starting on page 33. It's an easy job except for getting out the packing nut in a compression faucet. To get at the nut, gently chip away the wall's surface with a ball-peen hammer and small cold chisel and then grip the nut with a deep socket wrench *(right)*. Replace the faucet with a similar one; follow the manufacturer's instructions when installing.

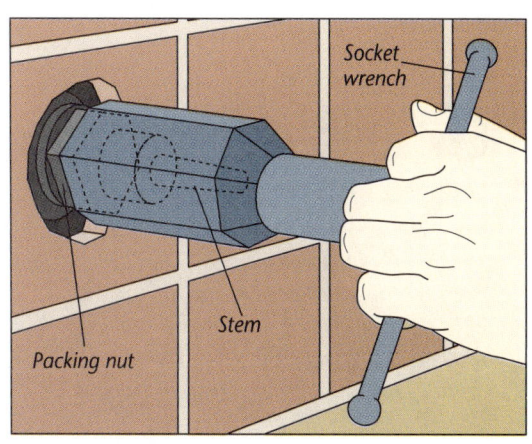

PLUMBING IMPROVEMENTS 79

OUTDOOR FAUCETS (HOSE BIBBS)

Several faucet models are available; nearly all have threaded spouts for attaching hoses. Some have bodies with female threads and are screwed onto a pipe; others have male threads and are screwed into a threaded T or elbow. One type fits onto horizontal pipes, another onto vertical pipes. One is designed for solvent welding to the end of a CPVC tube. Some hose bibbs have a notched flange, or escutcheon, allowing the faucet to be mounted on an exterior wall (this type is also called a sillcock).

First, decide where you want to install the new faucet. It should be convenient for outdoor watering and, if possible, high enough on the wall to clear a bucket. Be sure to consider, too, the location of the indoor cold water pipe you'll be tapping into (it's probably in your basement or crawl space). Plan carefully how you'll tap into the pipe and have ready all the pipe and fittings necessary before you begin. Consult the first chapter for more details on pipefitting.

Adding an outdoor faucet

TOOLKIT
- Drill and appropriate bit for wall material
- Hammer
- Screwdriver

1. Drilling through the wall
Before you start, check indoors to make sure you won't hit drainpipes, electrical lines, heating ductwork, studs, or floor joists. Avoid the foundation. If the water supply pipe is located below the foundation's top surface, plan to drill above the foundation and route the new pipes down to the water supply pipe. If possible, drill a small pilot hole from the inside out to mark the right location. Select the right bit for the job: a spade bit for wood, a masonry bit for brick or stucco. Then, using an extender if necessary, drill through the wall from the outside, boring a hole large enough to fit the pipe that will be attached to the faucet.

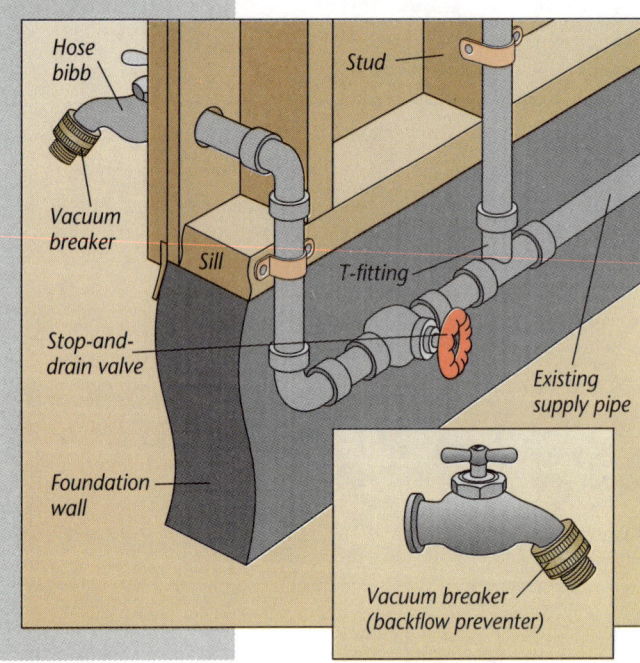

2. Connecting the faucet
Turn off the water at the house shutoff valve *(page 10)* and drain the pipes. Tap into the pipe with a T-fitting. You can add an indoor shutoff valve with a drain, called a stop-and-drain valve. This is a must in cold-winter areas, unless you're using a freezeproof faucet. The stop-and-drain valve allows the outside water to be turned off inside a warm basement or crawl space and the water beyond to be drained so it won't freeze.

Run new pipe through the wall *(left)* and connect the faucet to it. Connecting pipes must be well anchored to the house's framing (studs, joist, or sill) near the wall, as well as all along the pipe run. Fill any gaps around the pipe, both inside and out, with waterproof silicone rubber sealant or plastic foam sealant. When installing a flanged faucet, you can caulk the space around the pipe before screwing the flange in place.

Many codes require that you install a vacuum breaker or backflow preventer *(inset)*. This device, screwed on between the faucet spout and the hose itself, prevents the backflow of possibly polluted water into your home's water system.

 ASK A PRO

WHAT IF I LIVE IN AN AREA WITH COLD WINTERS?
If you live in an area where winter temperatures often dip below freezing, it makes sense to install a freezeproof faucet (right). This type of faucet has an elongated body that extends well into a basement or crawl space and a valve seat located far back into the body. When you turn off the faucet, the water flow stops back inside the house. Freezeproof faucets are self-draining. You install the unit at a slight tilt toward the ground outside, which lets any water remaining in the body after the faucet is turned off to run out.

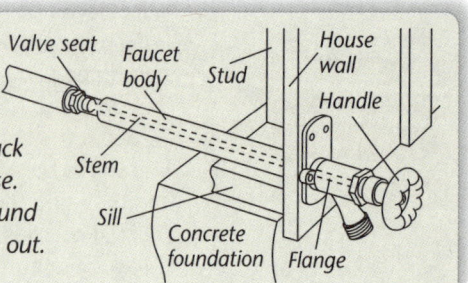

80 PLUMBING IMPROVEMENTS

PUTTING IN SINKS

Whether you're shopping for a wall-hung or deck-mounted sink, you'll have a choice of materials and styles: stainless or enameled steel, porcelain-coated cast iron, plastic, and vitreous china, available in single and double-well sinks in varying depths and shapes. All come with holes for either 4-, 6-, or 8-inch faucet assemblies. In most cases, it is much easier to hook up faucets and sink flanges before a sink is installed.

Wall-hung sinks are no longer as common for residential use as they used to be, but they can still be found in older homes. If you want to change from a wall-hung style to a deck-mounted, see the installation instructions on page 83. But if you're replacing a wall-hung with a new wall-hung model, consult the guidelines on this page for help.

A deck-mounted sink fits into a specially cut hole in a bathroom vanity or kitchen countertop. Whether frame-rimmed, self-rimmed, or unrimmed *(below)*, a deck-mounted sink is sealed with clamps or lugs and plumber's putty. For a deck-mounted sink, measure the hole in the countertop and take the measurements with you when you shop.

One variation of the deck-mounted sink that is increasingly popular is the one-piece molded sink with integral countertop. Easy to install, this type is set atop a cabinet and fastened from below.

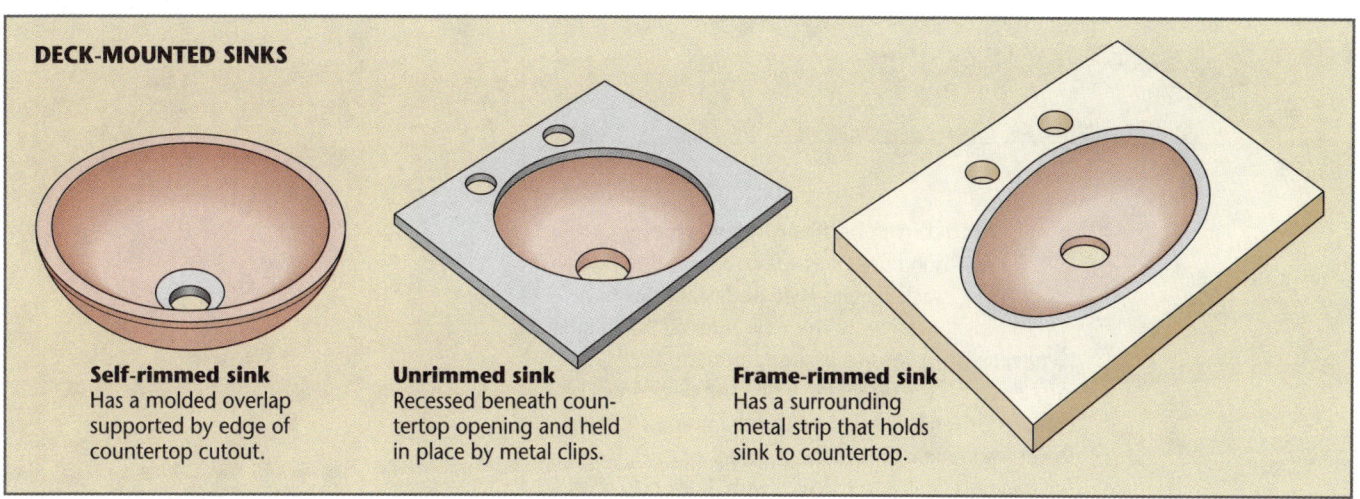

DECK-MOUNTED SINKS

Self-rimmed sink
Has a molded overlap supported by edge of countertop cutout.

Unrimmed sink
Recessed beneath countertop opening and held in place by metal clips.

Frame-rimmed sink
Has a surrounding metal strip that holds sink to countertop.

Replacing a wall-hung sink

TOOLKIT
- Basin wrench
- Adjustable wrench (optional)
- Compass or saber saw
- Chisel and mallet (optional)
- Hammer (optional)
- Screwdriver
- Carpenter's level
- Caulking gun

1 ▸ Removing the old sink
CAUTION: Before doing any work, turn off the water at the fixture shutoff valves or house shutoff valve *(page 10)*. Open the faucet to relieve the pressure.

Detach the faucet *(page 77)*; then disconnect the supply nuts at the faucet inlet shanks under the sink. Remove the trap from the tailpiece *(page 61)* and the drainbody tailpiece (it will come off as a unit). Pull straight up on the sink *(right)*. It should lift off its hanger, brackets, or pedestal; if not, check for hold-down bolts fastening it to the wall. Undo any bolts with an adjustable wrench. The pedestal that a wall-hung sink may be resting on is often bolted or grouted to the floor. Unfasten it, if necessary, and rock it back and forth to detach it.

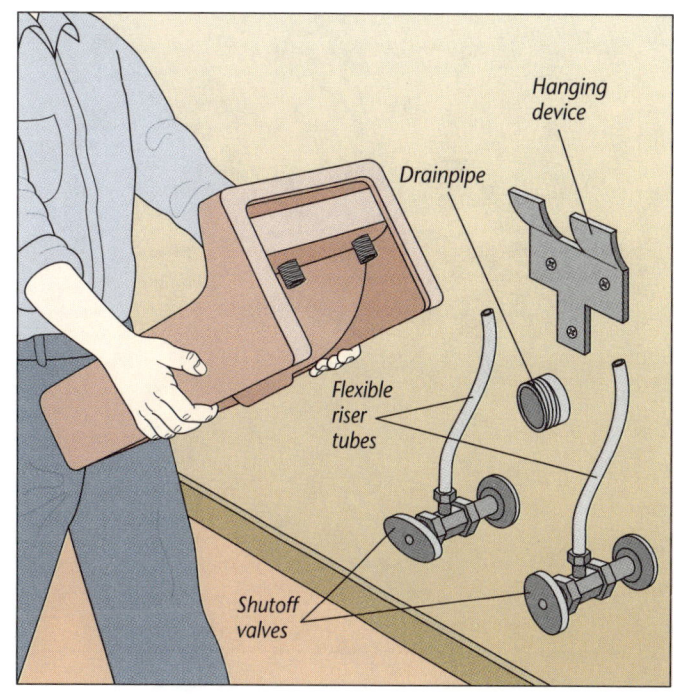

PLUMBING IMPROVEMENTS 81

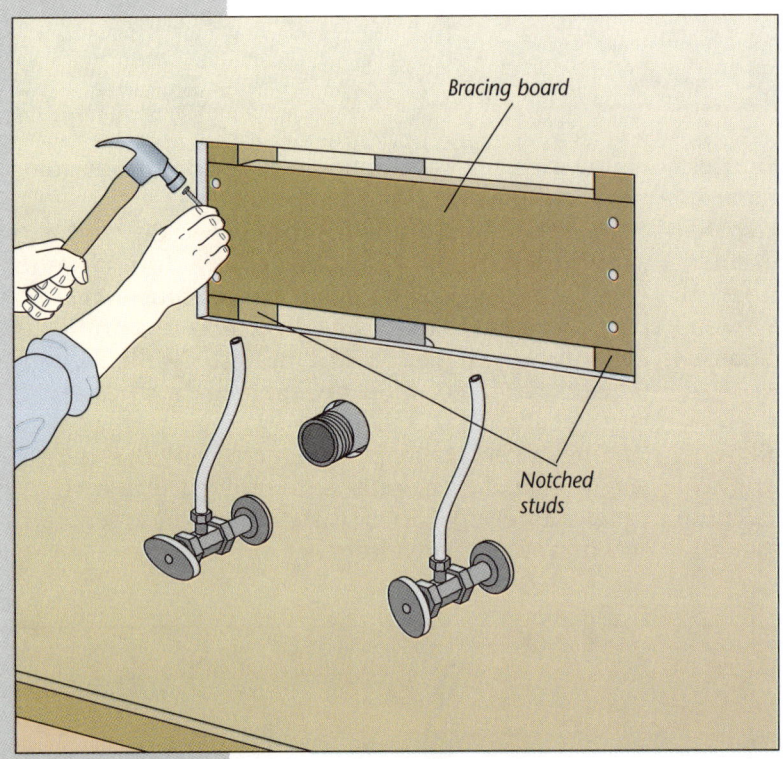

2. Installing a bracing board

For a first-time sink installation or for a wall-hung sink requiring extra support, you'll need to install a bracing board. Cut away the wall material between two studs to attach a 2x6 or 2x8 bracing board directly behind the sink. To get this board flush with the studs' surfaces, notch the studs (start with saw cuts; then chisel out the wood between them) and nail or screw the bracing board into place *(left)*; then finish the wall. Attach the sink hangers or brackets through the wall's surface.

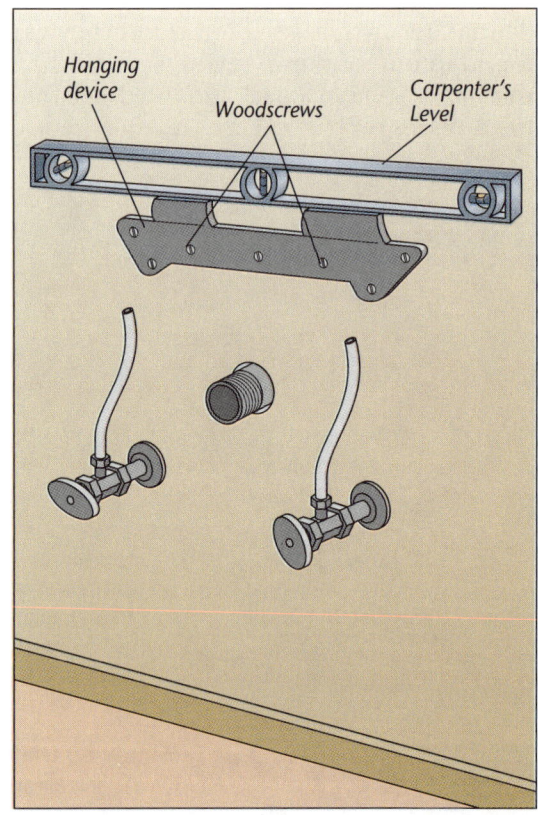

3. Attaching the hanging device

Wall-hung sinks come with supporting hangers or angle brackets. Old-style pedestal sinks have special clips that screw to the wall. Refer to the manufacturer's instructions to properly position the hanging device on the wall. Generally, the device will be centered, then leveled, over the drainpipe *(right)* at the desired height (31" to 38" above the floor). Fasten the hanging device to the wall with 3" woodscrews.

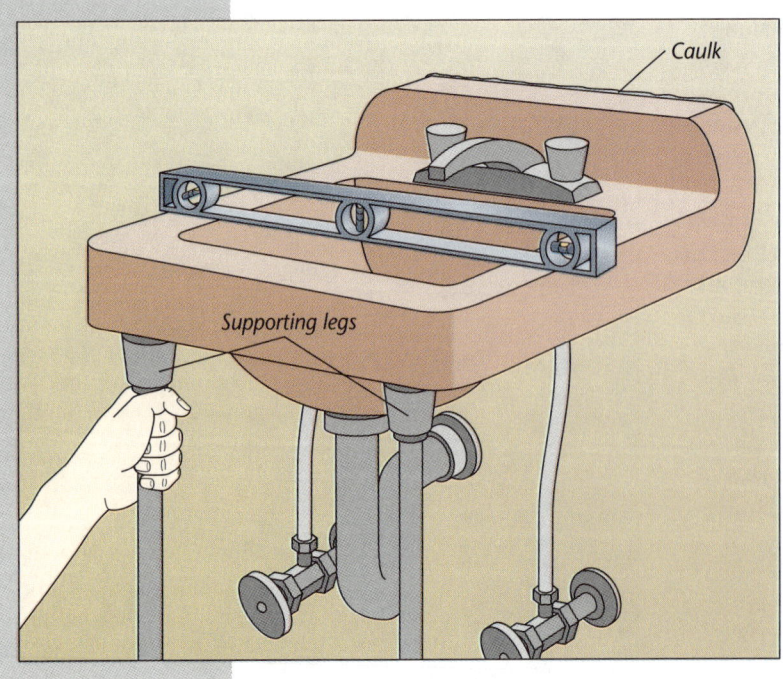

4. Reattaching the sink

Install the new tailpiece and drain body *(page 62)*, and riser tubes. Attach the faucet *(page 78)* and sink flange *(opposite page)*. Carefully lower the sink onto the hanger. Some hangers have projecting tabs that fit into slots under the sink's back edge. Angle brackets bolt up into the sink's base, but these can give way under downward pressure, so install front supporting legs on the sink to give it extra stability. Fasten the legs, then screw the adjusting section of each leg downward until the sink is level front and back *(left)*. Seal the sink-wall joint with a bead of plastic tub and tile caulk, or silicone rubber sealant. Hook up the supply pipes *(page 76)* and trap *(page 62)*, turn the water on, and tighten any leaky connections.

Installing a deck-mounted sink

TOOLKIT
- Screwdriver
- Saber saw
- Putty knife

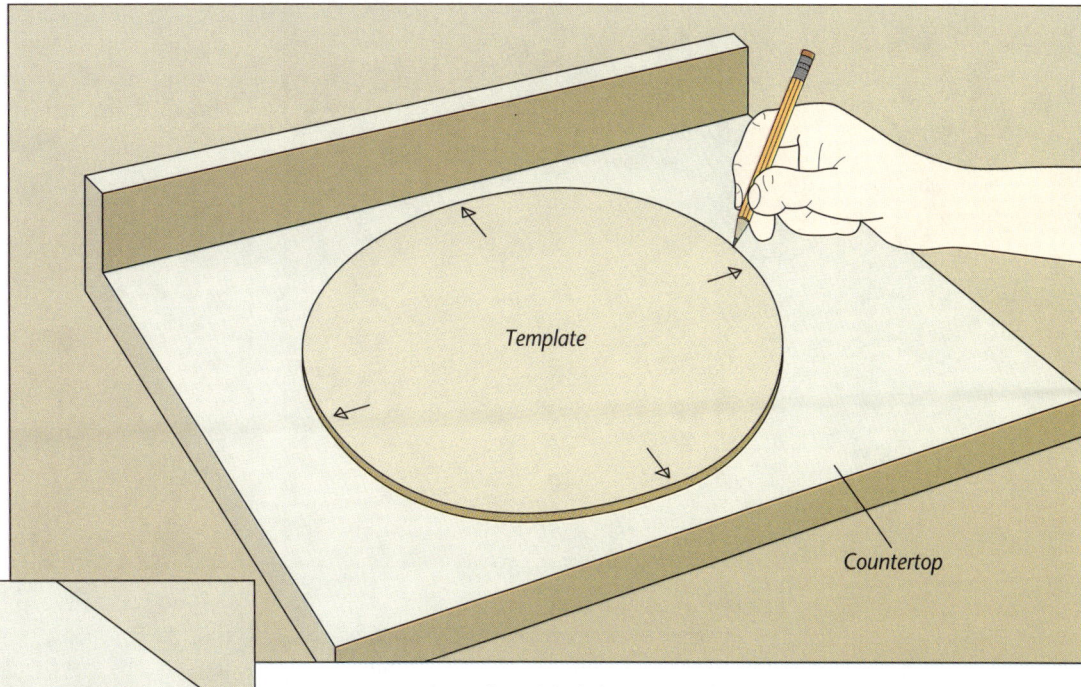

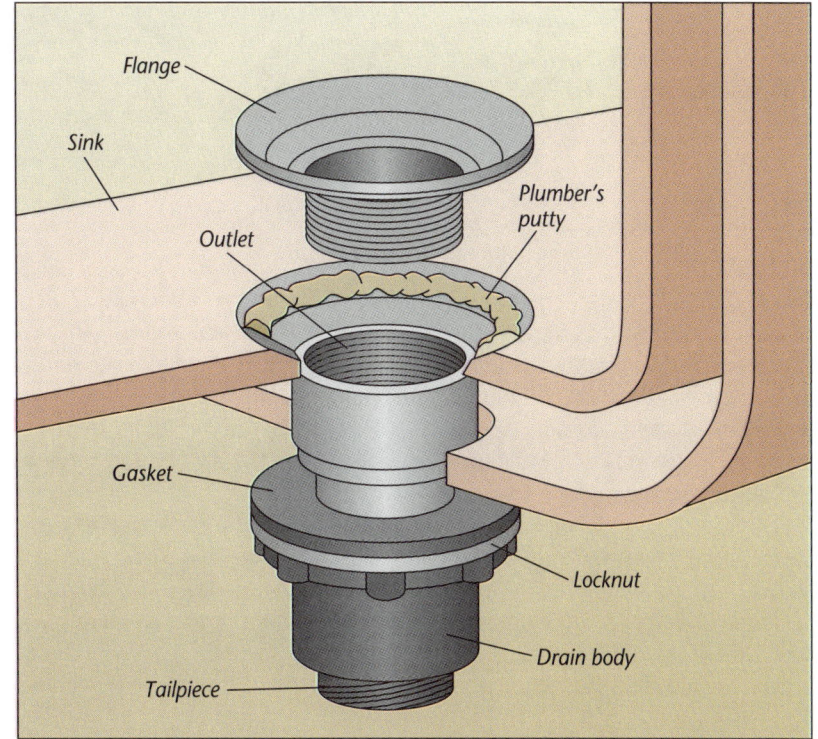

1 Removing the old sink and positioning the new one
First, shut off the water supply *(page 10)*; then drain and disconnect the water supply pipes *(page 76)* and the trap *(page 61)*. Remove a self-rimmed sink by forcing it free from below. With frame-rimmed and unrimmed sinks, you'll have to remove the clamps or lugs from below. CAUTION: Suspend the weight of the sink from above or find a helper to support it while you remove the last of the lugs. It should then lift right out.

If the sink is a new installation, first trace a template *(above)* or the bottom edge of the sink's frame *(left)* onto the exact spot on the countertop where the sink will sit. Use a saber saw to cut out the countertop opening. You'll find it's best to mount the faucet *(page 78)* and hook up the sink flange before you finally install the sink in the countertop.

2 Installing the flange
Kitchen sinks usually have strainers in their drains *(page 45)*; bathroom sinks may have pop-ups *(page 46)* and flanges. To install a flange, run a ring of plumber's putty around the water outlet. Then press the flange into the puttied outlet *(left)* and attach the gasket, locknut, and drain body to the bottom of the flange. Screw the tailpiece onto the drain body.

PLUMBING IMPROVEMENTS 83

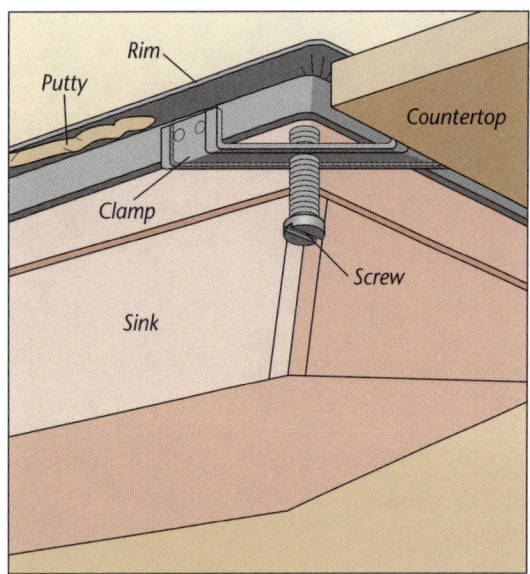
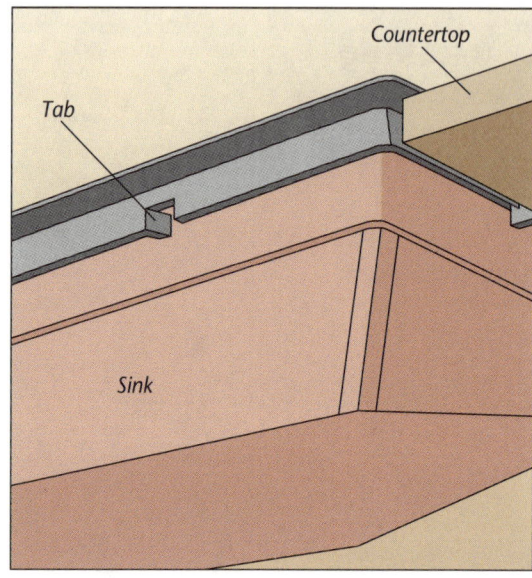

3 Framing the sink
For a frame-rimmed sink, apply a ring of plumber's putty around the top edge of the sink. Fasten the frame to the sink, following the manufacturer's instructions—some frames attach with metal corner clamps *(above, left)* or lugs, others with metal extension tabs that bend around the sink lip *(above, right)*. Wipe off excess putty. For self-rimmed sinks, place the putty between the sink and the rim.

4 Securing the sink
Before installing a deck-mounted sink of any style, apply a 1/2" wide strip of plumber's putty or silicone rubber sealant along the edge of the countertop opening. Set the sink into the hole, pressing it down. Smooth any excess putty. Anchor the sink at 6" to 8" intervals *(right)*, using any clamps or lugs provided. Hook up the supply pipes *(page 76)* and trap *(page 62)*. Turn on the water and check for leaks.

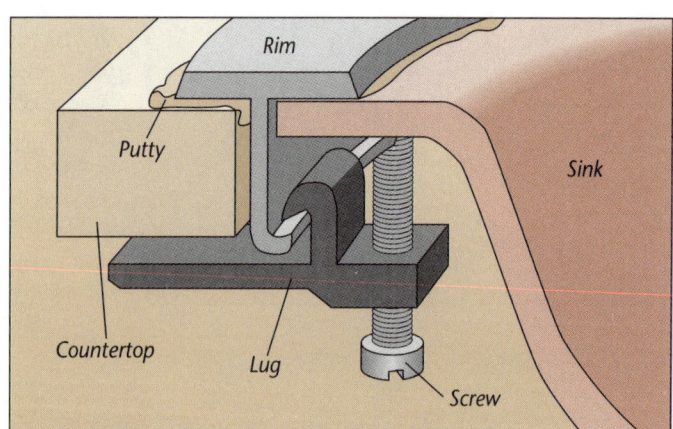

 MAINTENANCE TIP

CARING FOR NEW FIXTURES
Stainless steel fixtures stay shiny if given light and frequent cleaning with a damp cloth, followed by drying with a soft cloth. For a high polish, apply a mildly abrasive cleanser. For difficult spots and stains that aren't removed by daily cleaning, try ammonia in water, or use baking soda, vinegar, alcohol, or turpentine applied with a rag. Follow any of these applications with detergent and hot water; then rinse and dry with a soft, clean cloth.

Clean porcelain enamel fixtures gently with soapy water, then rinse and dry. For stubborn stains, fill the sink or tub with hot water and add chlorine bleach or oxygen-based bleach diluted according to label instructions. Let this stand and soak until the stain can be rubbed off. Full-strength or improperly diluted bleach can damage the surface. Abrasive cleansers can also do harm—they contain gritty substances that wear away the enamel surface, making it more difficult to clean as time goes on.

For toilets, use a chlorine bleach or a special toilet cleaner. Liquid cleansers or foaming spray bathroom cleaners are also available.

Fiberglass fixtures are coated with a protective gel sealant that wears off in time. Clean with liquid cleaners only. Abrasive cleansers should never be used on fiberglass—they will damage the surface. The original finish can be restored by the unit's manufacturer. Apply liquid automobile wax or fiberglass polish from time to time to maintain a protective shine.

84 PLUMBING IMPROVEMENTS

REPLACING A TOILET

If your toilet has seen better days, you'll be glad to know that replacing it is a one-afternoon project that you can tackle yourself. Installing a toilet in a new location is more of a challenge, because of the need to extend supply and drainpipes *(page 67)*. You may want to have a professional run the pipes to the desired spot, then do the installation yourself.

When shopping for a toilet, you'll find there are many choices—wall-mounted, bowl-mounted, water-saver, wash-down, reverse-trap, and siphon-jet models. The bowl-mounted type shown in this section is the most common.

Other than the code requirements for a new toilet, the only crucial dimension to consider when you're installing a toilet is its roughing-in size—the distance from the wall to the center of the drainpipe. You can usually determine roughing-in size without first removing the bowl—measure from the wall to the center of the two hold-down bolts that secure the bowl to the floor. (If the bowl has four hold-down bolts, you should measure to the rear pair.) The roughing-in distance for the new toilet can be shorter than that of the old fixture, but it cannot be longer or the new toilet won't fit.

Pick out a model that's ready to install—one that has a flush mechanism already in the tank. With the toilet, you'll get the necessary gaskets, washers, and hardware for fitting the tank to the bowl, but you may need to buy hold-down bolts and a wax gasket.

Also, buy a can of plumber's putty to secure the toilet base to the floor and the caps to the hold-down bolts. Finally, if the old toilet didn't have a shutoff valve, it's a good idea to install one now *(page 76)*.

Putting in a bowl-mounted toilet

TOOLKIT
- Rib-joint pliers
- Screwdriver and open-end wrench
- Adjustable wrench
- Fine-toothed hacksaw (optional)
- Putty knife
- Carpenter's level
- Spud wrench

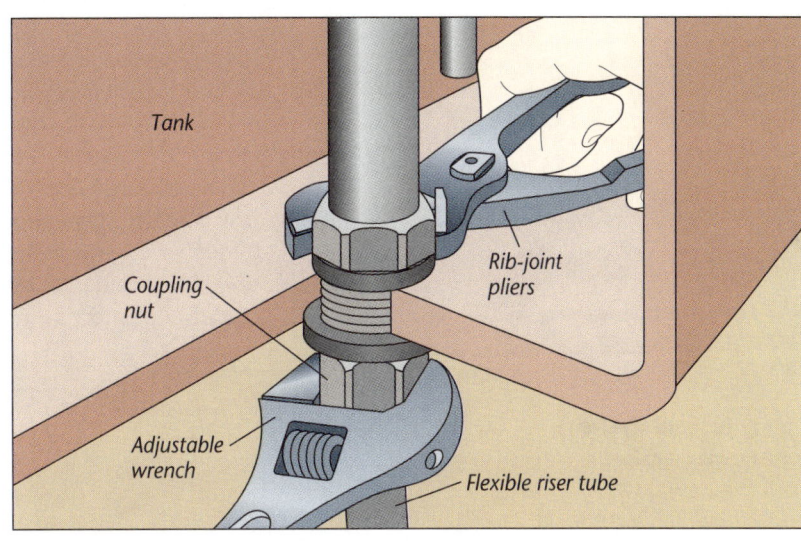

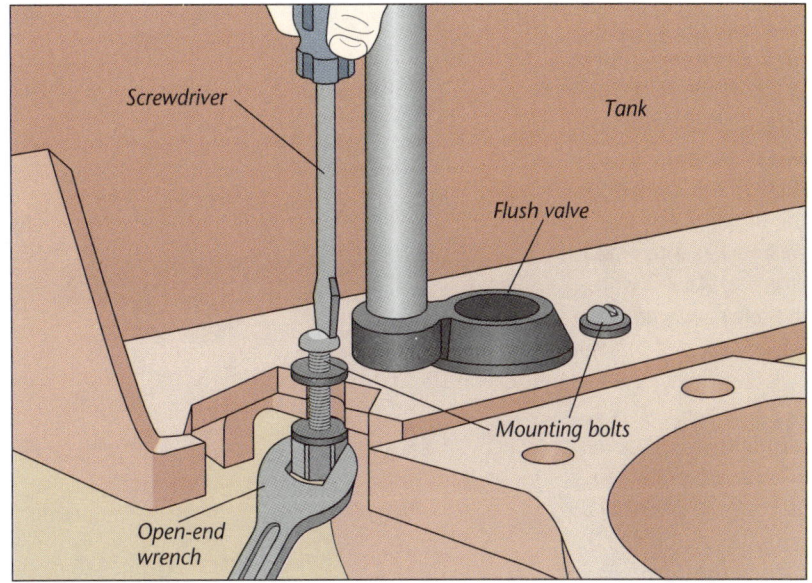

1. Removing the old tank

First, turn off the water at the fixture shutoff valve or house shutoff valve *(page 10)*. Flush the toilet twice to empty the bowl and tank. Sponge out any water that remains. Unfasten the coupling nut on the flexible riser tube *(left, above)* at the bottom of the tank. If the flexible riser tube is kinked or corroded, replace it with a new riser tube.

Unbolt the empty tank *(left, below)*, using a screwdriver to hold the mounting bolt inside the tank while unfastening its nut with a wrench from below.

PLUMBING IMPROVEMENTS 85

2 Removing the bowl
Pry the caps off the hold-down bolts and remove the nuts with an adjustable wrench. If the nuts are rusted on the bolts, soak them with penetrating oil or cut the bolts off with a fine-toothed hacksaw.

Gently rock the bowl from side to side to break the seal between the bowl and the floor. Lift the bowl straight up, tilting it forward slightly to avoid spilling any remaining water *(right)*. Stuff a rag into the drainpipe to minimize unpleasant sewage odors and to keep debris from falling into the opening.

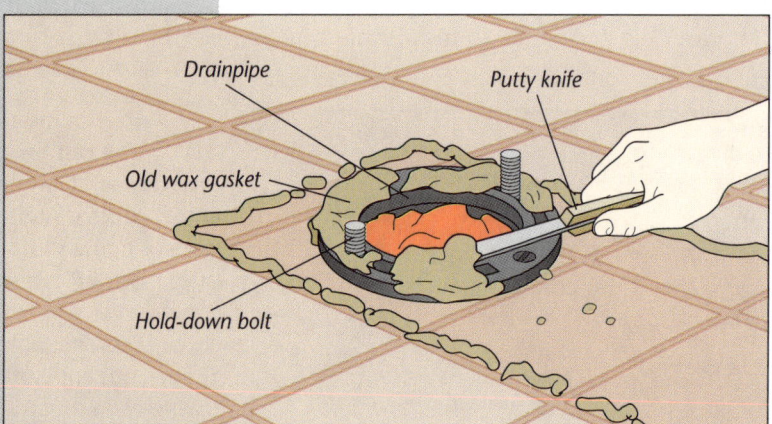

3 Preparing the floor flange
Scrape up the old wax gasket with a putty knife *(left)* and remove the old hold-down bolts from the floor flange. Thoroughly clean the flange to prevent leaks at the base of the new bowl, and check it for deterioration. If it's cracked or broken— or if you just want to guard against trouble later— replace the flange with a copper or plastic one that can be soldered or cemented into place. Set the new floor bolts in plumber's putty and insert them through the flange. Adjust the bolts so that they line up with the center of the drainpipe. Scrape up any residue of plumber's putty from the floor.

4 Installing the wax gasket
Turn the new bowl upside down on a cushioned surface. Place the new wax gasket over the toilet horn (outlet) on the bottom of the bowl *(right)*. The tapered side of the wax gasket should face away from the bowl. If you use a wax gasket with a plastic collar, install it with the collar away from the bowl. Make sure that the collar will fit into the toilet's floor flange. If not, substitute a wax gasket without a collar. In any case, apply plumber's putty to the bottom edge of the bowl.

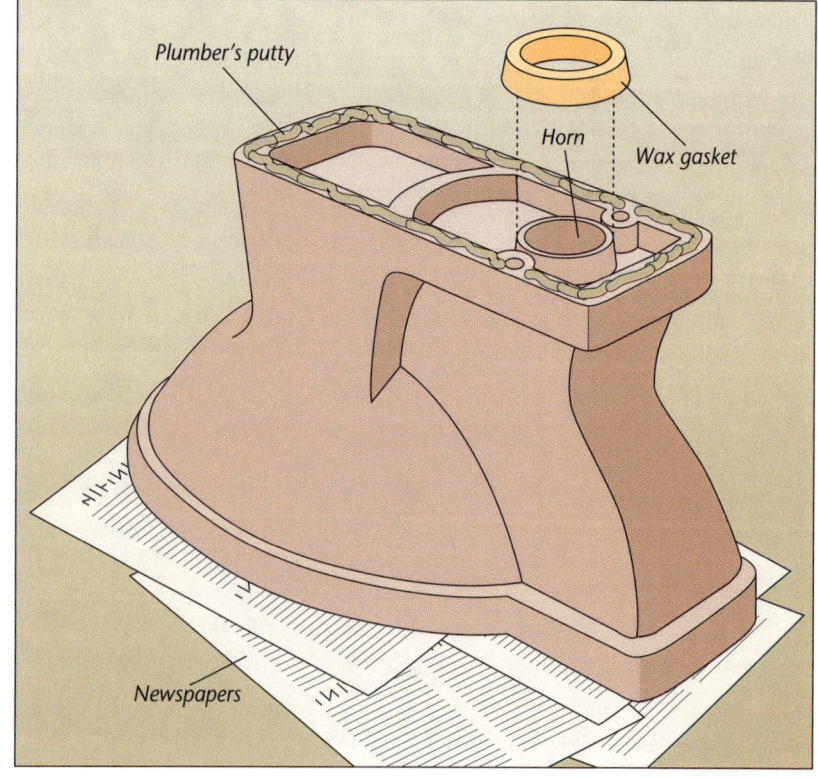

86   PLUMBING IMPROVEMENTS

5 ▶ Placing the bowl

Remove the rags from the drainpipe. Gently lower the bowl into place atop the flange, using the bolts as guides. Press down firmly, while twisting slightly and rocking.

Checking with a level, straighten the bowl *(right)*, and use thin pieces of metal to shim the bowl where necessary. Hand-tighten the washers and nuts onto the bolts.

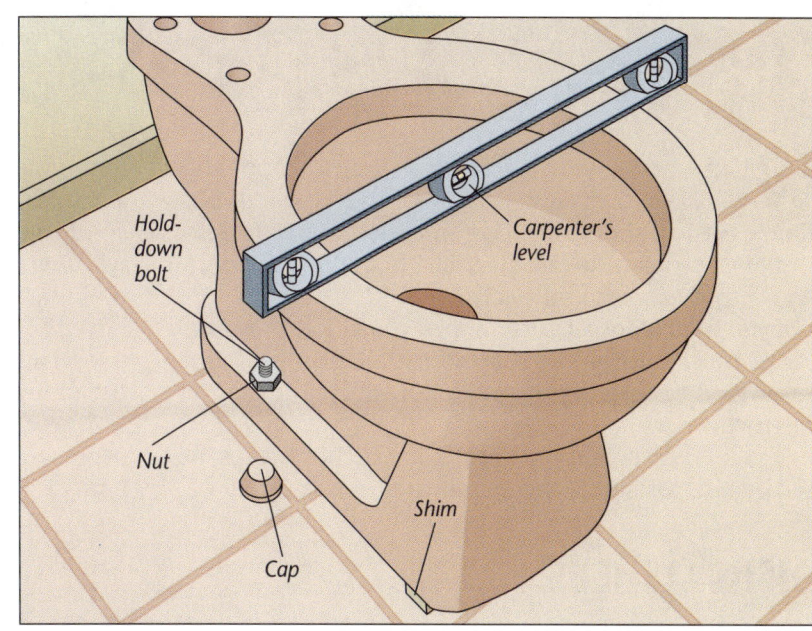

6 ▶ Attaching the tank

Fit the small rubber spud washer over the discharge tube and insert the tube through the flush-valve opening on the bottom of the tank. Thread the spud nut over the discharge tube and tighten it with a spud wrench. Then slip the bigger spud washer over the end of the discharge tube. Place the rubber tank cushion, if any, on the rear of the bowl. Position the tank over the bowl and tighten the nuts and washers onto the mounting bolts. Use an adjustable wrench to snug up the hold-down nuts at the base of the bowl, but don't over-tighten to avoid cracking the base. Check that the bowl is still level. Fill the caps with plumber's putty and place them over the bolt ends. Smooth the puttied joint at the base of the toilet bowl.

7 ▶ Hooking up the water supply

If your water supply stubout comes from the wall and the new tank is lower than the original, install an elbow fitting on the stubout. Use two 4" to 6" threaded nipples and a second elbow to connect the shutoff valve to the flexible riser tube *(right)*. Fasten the coupling nut on the flexible riser tube to the bottom of the tank.

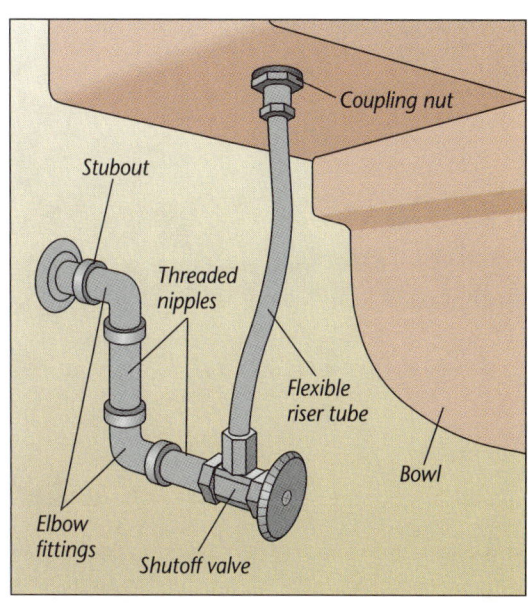

PLUMBING IMPROVEMENTS

INSTALLING APPLIANCES

When working with water-using electrical appliances you should always be aware of the danger of possible electrical shock. It's recommended that a professional electrician do the electrical hookups, unless you know your way around electricity.

If you want to install a dishwasher, you can do the plumbing portion yourself. You'll find installation guidelines below, but be sure to read and follow the manufacturer's instructions.

If you aren't able to compost your kitchen scraps, a garbage disposer may be the solution to getting rid of food quickly and easily. Details for setting up your new garbage disposer are on page 90.

Adding the necessary plumbing connections for a washing machine is a fairly straightforward task. Most building codes now require that new residential units have the connecting pipes already installed, but if you live in a home that's never had a washer, you can do the plumbing yourself *(page 91)*.

A new water heater may be needed if your old one begins to leak or show signs of rust and corrosion. To replace your water heater, turn to page 92.

DISHWASHERS

In terms of energy-saving, dishwashers are available with built-in preheating elements that independently heat the water to 160° so that the rest of the plumbing system can operate with cooler water.

A built-in dishwasher needs a hot water supply pipe connection (cold isn't necessary), a drainpipe fitting, and an air-gap venting hookup. Some areas require a permit and an inspection for a built-in dishwasher.

If the dishwasher has already been wired in, even though the power has been turned off, make sure you don't touch any bare wire ends with your hands or with tools; handle them by the insulation only. You should test all exposed wires to each other and to ground with a voltage tester to make certain the circuit is dead before proceeding further. The voltage tester should not glow on any test.

Hooking up a dishwasher

TOOLKIT
- Voltage tester
- Fine-toothed hacksaw
- Adjustable wrench
- Screwdriver
- Hammer and old screwdriver
- Drill with a hole saw
- Spud wrench
- Carpenter's level

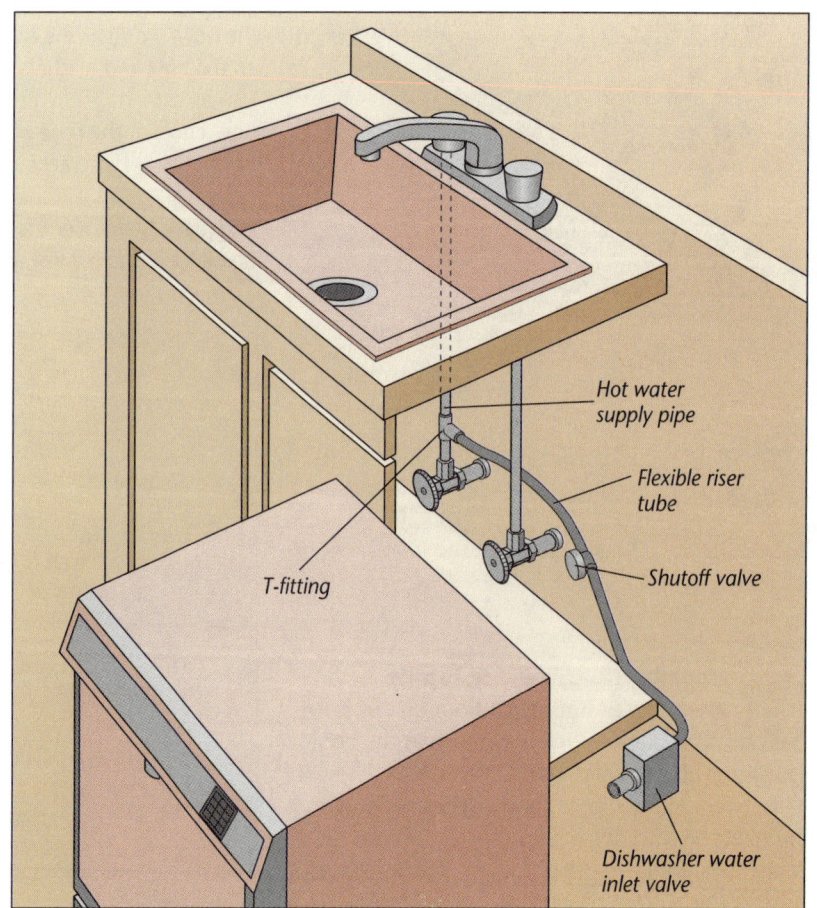

1 Connecting to the supply pipe
Begin by shutting off the water supply *(page 10)* and draining the hot water supply pipe that you plan to tap into. Cut into the pipe and install a T-fitting, as described on page 71, or a special three-way valve to isolate the water supply to the dishwasher.

Run a flexible riser tube from the T-fitting *(left)* to the water inlet valve for the dishwasher. Always install a shutoff valve *(page 76)* in the dishwasher supply pipe. Many codes require a dishwasher's inlet to be connected with a backflow-prevention device; even if your code does not require this, it's a good idea anyway.

88 PLUMBING IMPROVEMENTS

2. Setting up the drainage system (sink trap or garbage disposer)

Your dishwasher can drain either into the sink trap or into a garbage disposer. For use with a sink drain, you'll need to buy a tapped waste T-fitting *(below, left)* or tapped tailpiece fitting.

To install the fitting, remove the sink tailpiece *(page 62)* and insert the new fitting into the trap. Secure it by tightening the slip nut on the trap. Cut the tailpiece to fit between the sink strainer and fitting or trap. Reattach the tailpiece and clamp the dishwasher drain hose to the fitting. You may need a rubber adapter, which fits between any size of dishwasher hose and the fitting. It tightens over the hose and fitting with screw clamps.

The dishwasher drain hose can be attached to a garbage disposer instead *(below, right)*. Turn off the electrical circuit to the disposer; use a hammer and an old screwdriver to punch out the knockout plug inside the disposer's dishwasher drain fitting. Clamp the dishwasher drain hose to the fitting.

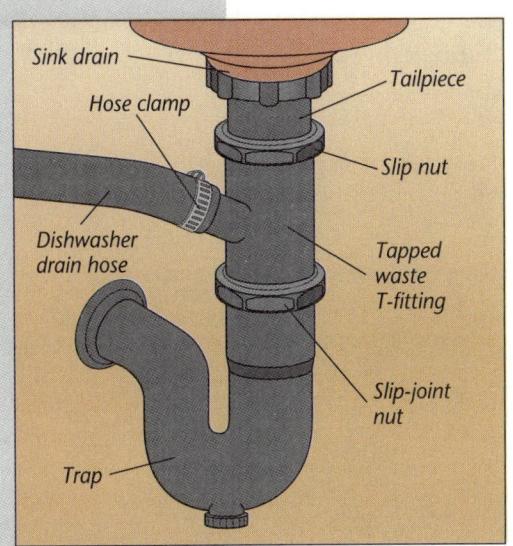

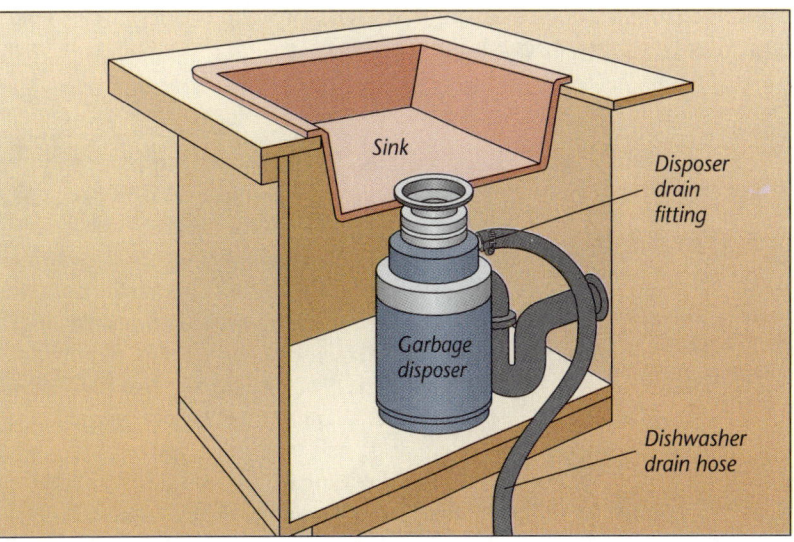

3. Adding on an air gap

Most codes require you to connect an air gap *(left)* to the dishwasher's drain hose to prevent contamination of the potable water supply system. Where this is not required, you can make a gradual loop with the drain hose to the height of the dishwasher's top instead. However, it's a good idea to use the air gap.

To install the air gap, first bore a hole into the countertop using an electric drill with a hole saw installed. Mount the air gap by inserting it through the hole and secure it by tightening the locknut (found underneath the device) with a spud wrench. Following manufacturer's instructions, make the connections to the dishwasher drain hose and sink trap.

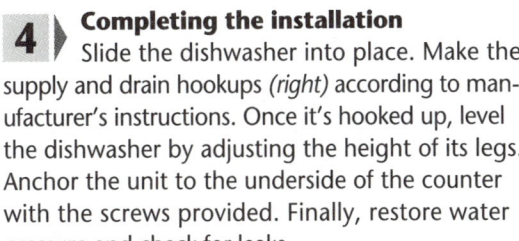

4. Completing the installation

Slide the dishwasher into place. Make the supply and drain hookups *(right)* according to manufacturer's instructions. Once it's hooked up, level the dishwasher by adjusting the height of its legs. Anchor the unit to the underside of the counter with the screws provided. Finally, restore water pressure and check for leaks.

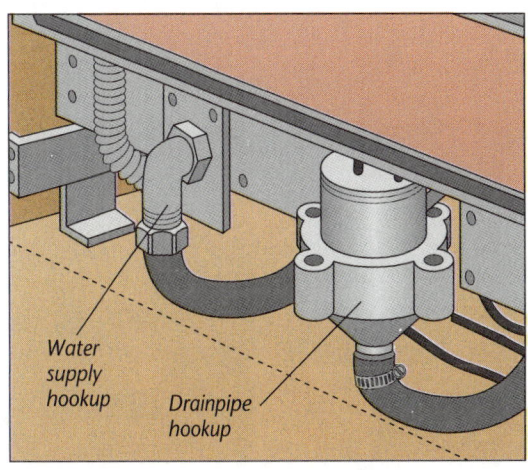

PLUMBING IMPROVEMENTS

GARBAGE DISPOSERS

Before you install a garbage disposer, check the codes in your area for any restrictions. Installation takes only a few hours, and the work isn't very difficult. Most units fit the standard 3½- or 4-inch drain outlets of kitchen sinks and are mounted somewhat like a sink strainer. However, like all plumbing products, garbage disposers vary from brand to brand.

Plumbing a disposer means altering the sink trap to fit the unit. Some models have direct wiring that should be connected by a licensed electrician. Other plug-in disposers require a 120-volt grounded GFCI-protected outlet under the sink—another job for an expert.

Always be cautious—water and watts don't mix. If you're replacing a disposer, turn off the electricity to that circuit and unplug the unit or disconnect the wiring before removing it. If the disposer is wired in, and even if the power is off, don't touch bare wire ends with hands or tools; handle them by the insulation only. Test exposed wires to each other and to ground with a voltage tester to be certain the circuit is dead before you proceed. The tester should not glow on any test.

You'll get detailed installation instructions with your disposer; follow them. Below are the typical steps to follow when you hook one up.

Installing a garbage disposer

TOOLKIT
- Voltage tester
- Putty knife
- Screwdriver
- Adjustable wrench
- Fine-toothed hacksaw (optional)

1 ▸ Putting in the mounting assembly

Disconnect the tailpiece and trap *(page 61)* from the sink strainer. Disassemble the sink strainer *(page 45)* and lift it out of the sink. Clean away any old putty or sealing gaskets from around the opening.

The disposer will come with its own sink flange and mounting assembly. Run a rim of plumber's putty around the sink opening and seat the flange in it. Then, working from below, slip the gasket, mounting rings, and snap ring *(right)* up onto the neck of the sink flange. The snap ring should fit firmly into a groove on the disposer's sink flange to hold things in place temporarily.

Uniformly tighten the slotted screws in the mounting rings until the gasket fits snugly against the bottom of the flange. Remove any excess putty from around the flange.

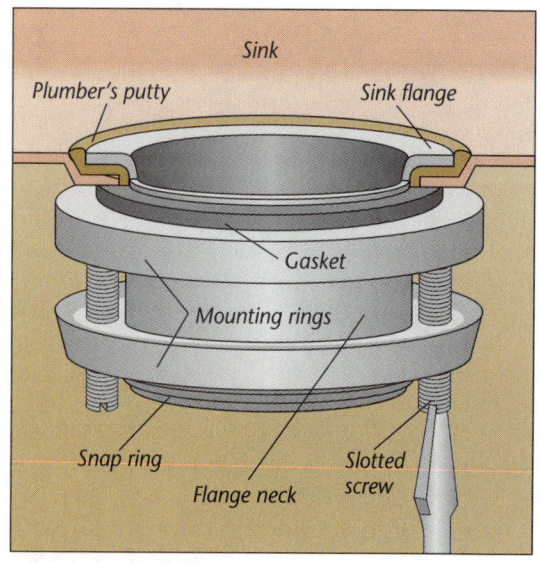

2 ◂ Fastening the unit

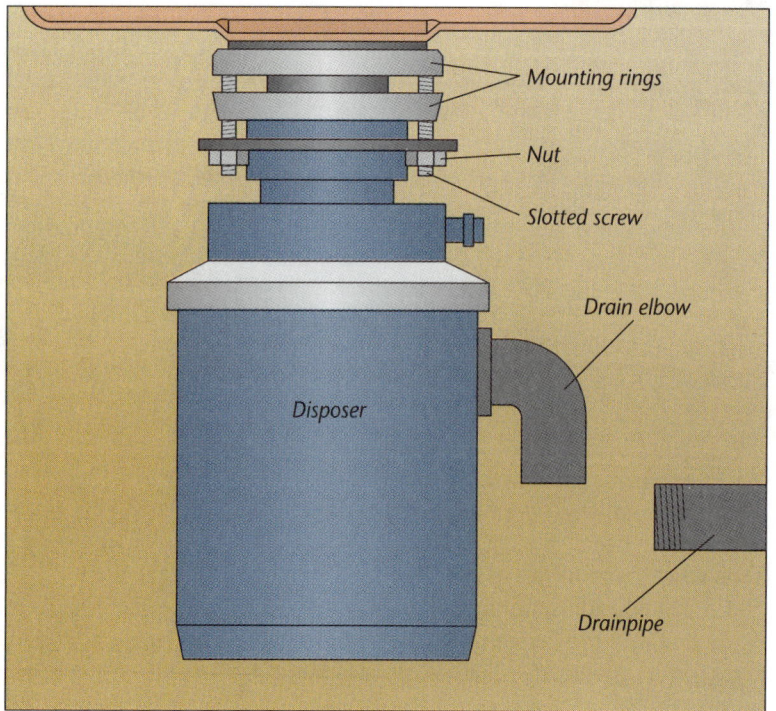

Attach the drain elbow to the unit, following the manufacturer's instructions. Align the holes in the unit's flange with the slotted screws in the mounting rings. Rotate the disposer so that the drain elbow lines up with the drainpipe. Then, tighten the nuts securely onto the slotted screws to ensure a good seal.

90 PLUMBING IMPROVEMENTS

3 **Hooking up the drain**
Fit one of the slip nuts and a washer onto the drain elbow *(right)*, then fasten the trap to the drain elbow. Add an elbow fitting onto the other end of the trap to adapt to the drainpipe. You may need to cut the elbow to make the connection. Run water down through the disposer to test for leaks. Tighten any loose connections.

The last part of the installation requires electrical know-how. Call in an electrician to run the wires from a power source to an outlet under the sink and to the disposer's ON/OFF switch.

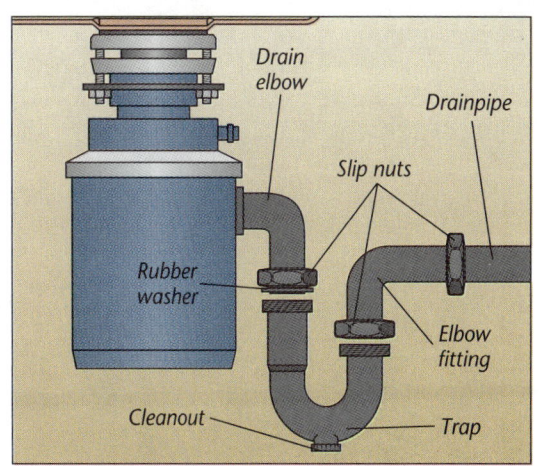

WASHING MACHINES

You'll need to run both hot and cold water supply pipes to the desired location. In addition, each supply pipe needs a shutoff valve *(page 76)* and, to prevent banging, a water hammer arrester. Install the latter according to the manufacturer's instructions.

First, locate and drain the nearest hot and cold water pipes. Supply pipes for an automatic washer are usually 1/2 inch in diameter. Check your local code and also the manufacturer's instructions before you install supply pipes. Extend the pipes *(page 67)* to the desired point just above the washer and install a T-fitting at the end of each pipe. Refer to the first chapter for instructions on pipefitting.

If there's no sink or laundry tub nearby, you'll need to drain the washer into a standpipe—a 2-inch-diameter pipe with a built-in trap that taps into the nearest drainpipe. If there is a sink, the drain hose is designed to hook over the edge of the laundry tub.

Connecting a washing machine

TOOLKIT
- Appropriate tools for type of pipefitting
- Screwdriver
- Fine-toothed hacksaw

1 **Installing shutoff valves**
Extend the pipes from the T-fittings, leaving enough space above the washing machine for shutoff valves. Install either two hose-bibb valves or a single-lever valve; buy the appropriate model for your type of pipe. You should close the shutoff valves when the machine is not in use. This relieves the constant pressure on the supply hoses and the water inlet valve—and could prevent a flood.

Hose-bibb shutoff valves *(below, left)* are often used. To install, add elbows at the end of the supply pipes and attach threaded nipples, then hose-bibbs to accept the machine hoses.

A single-lever shutoff valve *(below, right)* turns off both hot and cold water simultaneously. It can replace existing valves with little or no modification. Unscrew the valve adapters from the single-lever unit and attach one to the end of each supply tube; refer to the section starting on page 11 for pipe-fitting instructions. Slide the gaskets and valve body onto the valve adapters. Insert and tighten the attachment screws, then thread on the washer hoses. Many codes require that a washing machine's inlet be connected through a backflow-prevention device, which is a good practice anyway.

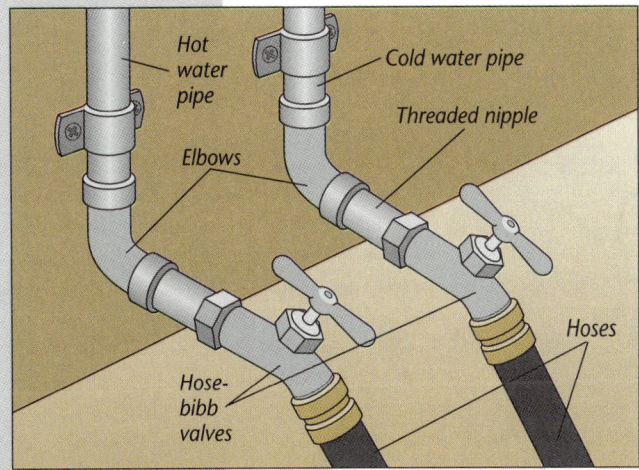
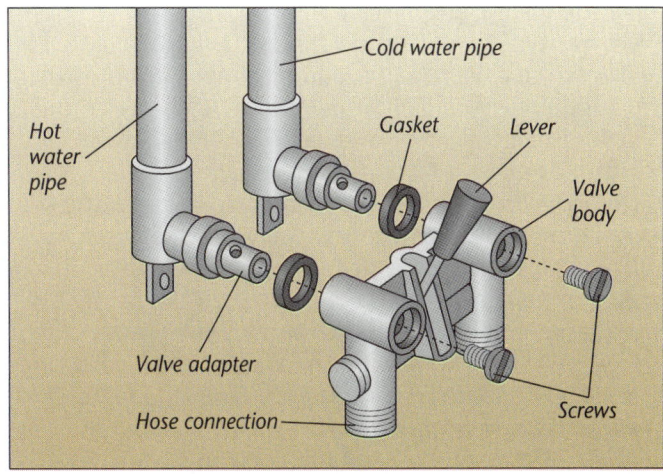

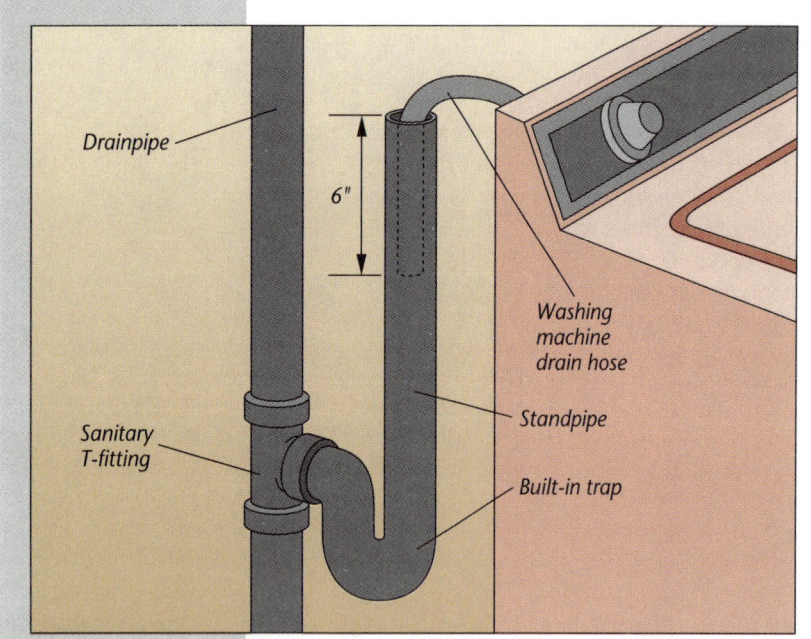

2 Attaching the standpipe
The standpipe, available in lengths from 34" to 72", should be taller than the highest level of water in the washer to prevent backup and siphoning of dirty water into the machine. To determine the size of standpipe you'll need, check the manufacturer's instructions. To install a standpipe, cut into a drainpipe and install a sanitary T-fitting. Attach the standpipe to the T-fitting and push the washing machine's drain hose down into the standpipe about 6"—be sure the hose won't be forced out of the pipe by the water pressure. As with other wet-vented drains, the standpipe's pipe must enter the drainpipe above the highest fixture drain. If this can't be done simply, its trap seal can be protected by an antisiphon vent *(page 70)*.

WATER HEATERS

When shopping for a new water heater, consider its capacity, recovery rate, warranty, tank lining, and fuel source.

Capacity: Gas heaters are usually sized from 30 to 80 gallons; electric heaters, because they are slower to recover, hold up to 102 gallons. The graphs shown below will give you a rough idea of the size of water heater tank that you need. The capacity should be based on the number of people and bathrooms in the household. Recovery rate and the incoming water temperature are also factors. The graphs are based on households with two water-using appliances; a household without a dishwasher or washing machine will need a smaller tank. There are exceptions; for families with small children, for example, more hot water is required for laundry.

Recovery rate: This refers to the number of gallons per hour that a heater can raise to 100°. Gas heaters have the fastest rate.

Warranty: Most water heaters come with a 7- to 15-year warranty. While it's good to choose a top-of-the-line model, beware: Some "deluxe energy-saver" models cost significantly more and carry a longer warranty, but don't necessarily cost less to operate.

Lining: A lined tank is recommended. Glass is the most common lining for water heater tanks; it resists corrosion and provides cleaner water than other linings. Copper-lined tanks are more durable than glass-lined tanks but are also much more expensive.

Fuel: It's easier to replace the water heater with one that uses the same type of fuel. In a new installation, availability and cost of gas versus electricity should be your primary consideration. You can start saving energy in your home by making adjustments to your water heater. Most heaters now come with 2 inches of built-in foam insulation—more efficient than fiberglass, which it is replacing. Further insulate the tank by wrapping it in fiberglass or foam (for savings of up to 10 percent on your water heating bill); kits are available or you can make your own. CAUTION: On a gas heater, keep insulation away from the pilot light and flue pipe; don't cover the top.

Lowering the temperature setting on your water heater to about 110° or 120° will help you to save fuel without making a noticeable difference in laundry or bathing. Turn down the heater during the periods when your house will be empty, or install an automatic timer which will adjust the thermostat during low- and peak-usage times.

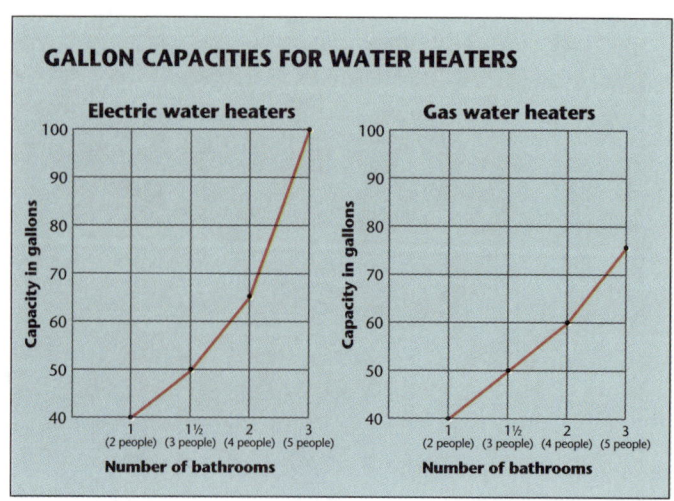

Water heaters use 240-volt electricity, so it's best to have an electrician handle the electrical work in removing and replacing an electric water heater. If you need to run gas piping to a new water heater, call a professional to do the work.

The temperature and pressure relief valve should be installed in its 3/4-inch side tapping and connected to a properly plumbed pressure-relief line. To install a relief line, use rigid pipe (metal or plastic) the same size as the relief valve. It shouldn't have any restrictions, and should extend—without dips, which could collect water and freeze—to within 6 inches of a floor drain or to a good drain spot outdoors. Check with your local plumbing inspector.

Replacing a water heater

TOOLKIT
- 2 pipe wrenches
- Fine-toothed hacksaw (optional)
- Neon tester (for electric heaters)
- Carpenter's level
- Screwdriver
- Adjustable wrench

1 Disconnecting the old unit

When you're ready to replace your water heater, first shut off the water and fuel (or power) supply to the old unit. If there's no floor drain beneath the valve, connect a hose to the drain valve near the base of the tank and run it to a nearby drain or outdoors. Then drain all the water out of the heater storage tank by opening the drain valve. Next, disconnect the water inlet and outlet pipes from the heater. If they are joined by unions *(right)* or flexible pipe connectors, use 2 pipe wrenches to unscrew them. If not, cut through the pipes with a fine-toothed hacksaw.

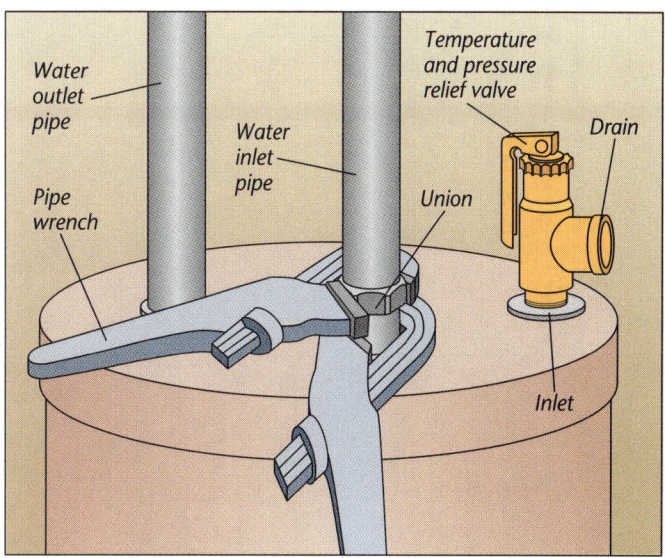

2 Detaching the power or fuel line

Unless you know what you're doing, the electrical work is best done by an electrician. To disconnect the power supply lines on an electric heater, shut off the power at the two circuit breakers and test exposed wiring with a neon tester between each wire and a good ground, such as metal piping or the water heater itself. Don't touch exposed wires with bare hands or tools; handle them only by their insulation. When you're sure the power has been shut off, remove the electrical cable from the heater *(below, left)*. For a gas or oil water heater, shut off the fuel and use pipe wrenches to disconnect the union between the fuel supply pipe and the inlet valve *(below, right)*. You'll also need to unscrew and remove the flue hat from the flue pipe of a gas heater. All water heaters should have temperature and pressure relief valves to prevent explosions in case the heating mechanism fails. The valves are inexpensive; get a new one when you get a new heater. You may be able to use the existing relief pipe *(left)*.

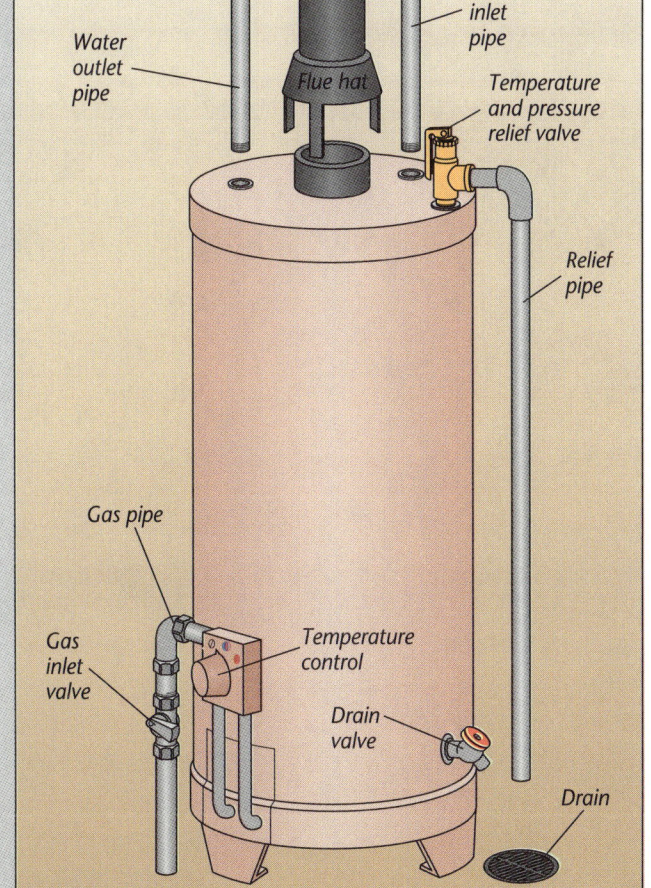

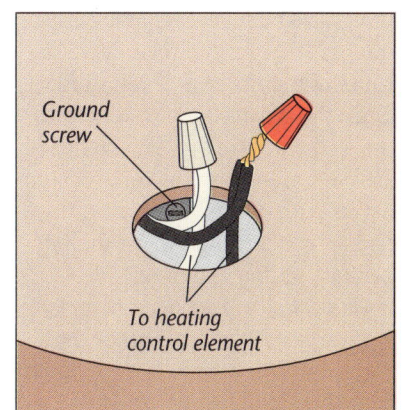

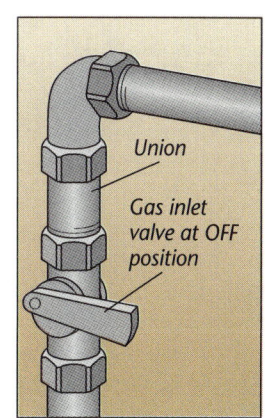

PLUMBING IMPROVEMENTS

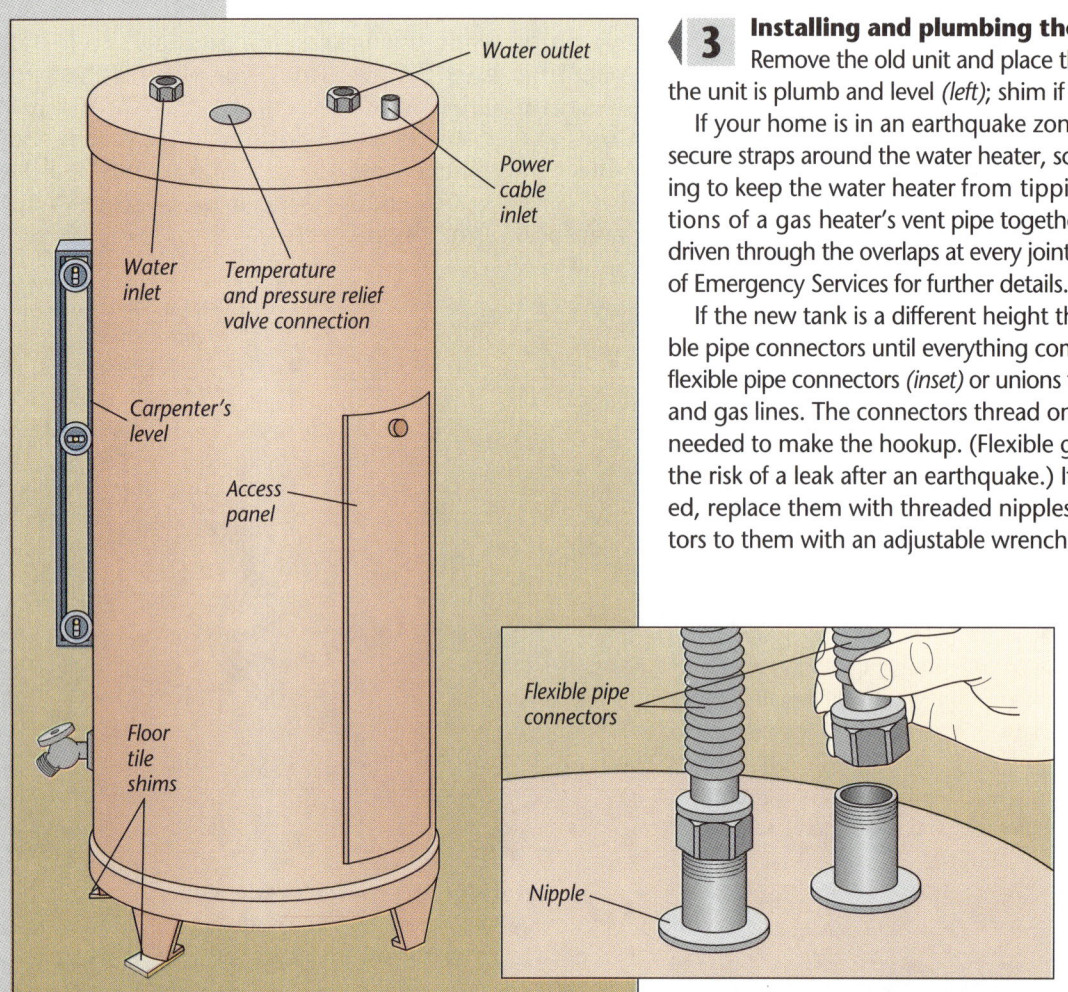

3 Installing and plumbing the new unit

Remove the old unit and place the new one. Check that the unit is plumb and level *(left)*; shim if necessary with floor tiles.

If your home is in an earthquake zone, you'll need to install secure straps around the water heater, screwing them to the framing to keep the water heater from tipping over. Fasten the sections of a gas heater's vent pipe together with sheet metal screws driven through the overlaps at every joint. Contact your state Office of Emergency Services for further details.

If the new tank is a different height than the old one, use flexible pipe connectors until everything comes together properly. Use flexible pipe connectors *(inset)* or unions to hook up both the water and gas lines. The connectors thread onto the pipe and bend as needed to make the hookup. (Flexible gas supply line can reduce the risk of a leak after an earthquake.) If the pipes aren't threaded, replace them with threaded nipples and secure the connectors to them with an adjustable wrench.

4 Activating the heater

For an electric heater, have an electrician connect the wiring for you. With all the connections made, open the water inlet valve to the heater. When the tank is filled with water, bleed the supply pipes by opening the hot water faucets to allow air to flow out of the pipes.

Test the temperature and pressure relief valve by squeezing its lever. Do not cap the pressure relief valve (because of the risk of an explosion). Open the gas inlet valve or energize the electrical circuit to fuel the heater. For gas heaters, light the pilot according to instructions (usually on the control panel plate). Adjust the temperature setting as desired.

Finally, check all connections for leaks. If you're working on a gas heater, brush soapy water on the connections *(right)* —bubbles indicate a gas leak.

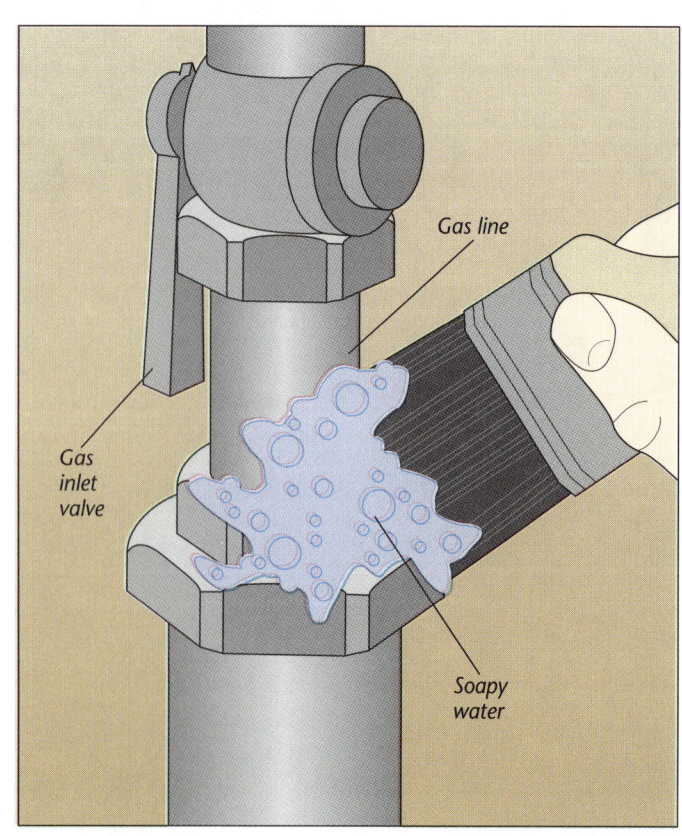

94 PLUMBING IMPROVEMENTS

PLUMBING GLOSSARY

ABS
Acrylonitrile-butadiene-styrene; rigid plastic drainpipe.

Adapter
Connects one type of pipe to another.

Cap
Fitting with a solid end used for closing off a pipe.

Center-to-center
In mounting faucets: Distance between centers of holes on a sink deck. In pipefitting: Distance between centers of two consecutive pipes.

Cleanout
Opening providing access to a drainline or trap; closed with a threaded cleanout plug.

Compression fitting
Easy-to-use fitting for copper or plastic tube. Pushed in and hand-tightened.

Compression nut
Used with a compression ring to join a flexible tube to a compression fitting.

Coupling
Fitting used to connect two lengths of pipe in a straight run.

Coupling nut
Holds a supply tube to a faucet inlet or a toilet inlet valve.

CPVC
Chlorinated polyvinyl chloride; rigid plastic tube for hot and cold water.

Critical distance
Maximum horizontal distance allowed between a fixture trap and a vent or soil stack.

Cross connection
Plumbing connection that could mix contaminated water with potable water supply.

DWV
Drain-waste and vent; system that carries away waste water and solid waste, allows sewer gases to escape, and maintains atmospheric pressure in drainpipes.

Elbow
Fitting used for making turns in pipe runs (for example, a 90° elbow makes a right-angle turn). A street elbow has one male and one female end.

Escutcheon
Decorative trim piece that fits over a faucet body or pipe extending from a wall.

Female
Pipes, valves, or fittings with internal threads.

Fitting
Device used to join pipes.

Fixture
A non-powered water-using device such as a sink, bathtub, shower or toilet.

Flange
Flat fitting or integral edging with holes to permit bolting together (a toilet bowl is bolted to a floor flange) or fastening to another surface (a tub is fastened to wall through an integral flange).

Flare fitting
Threaded fitting used on copper and plastic pipe that requires enlarging one end of the pipe.

Flexible connector
Bendable piece of tubing that delivers water from a shutoff valve to a fixture or appliance.

Flue
Large pipe through which fumes escape from a gas water heater.

Gasket
Device (usually rubber) used to make a joint between two parts watertight. Term is sometimes used interchangeably with washer.

Hose bibb
Valve with an external threaded outlet for accepting a hose fitting.

Joist
A horizontal wood framing member placed on edge, as a floor or ceiling joist.

Locknut
Nut used to secure a part, such as a toilet water inlet valve, in place.

Male
Pipes, valves, or fittings with external threads.

Nipple
Short piece of pipe with male threads used to join two fittings.

No-hub
Cast-iron pipe joined with neoprene gaskets and clamps.

O-ring
Narrow rubber ring; used in some faucets as packing to prevent leaking around stem and in swivel-spout faucets to prevent leaking at base of spout.

Packing
Material that stops leaking around the stem of a faucet or valve.

PB
Polybutylene; flexible plastic tubing for hot or cold water.

PE
Polyethylene; flexible plastic tubing for cold water outdoors.

Pipe
Drain pipe or supply pipe that is sized nominally by iron pipe sizes.

Pipe-joint compound
Sealing compound used on threaded fittings (applied to external threads).

Pipe-thread tape
Special tape used as a joint sealer in place of pipe joint compound.

Plug
Externally-threaded fitting for closing off a fitting that has internal threads.

PP
Polypropylene; rigid plastic pipe used for traps.

Pressure regulator
Device installed in a water supply line to reduce water pressure.

PVC
Polyvinyl chloride; rigid plastic pipe for cold water outdoors. Also, off-white piping used for DWV systems.

Reducer
Fitting that connects pipe of one diameter with pipe of a smaller diameter.

Riser
Vertical run of pipe.

Run
Horizontal or vertical series of pipes.

Saddle tee
T-fitting that is fastened onto side of pipe, eliminating cutting and threading or soldering; usually requires drilling into pipe.

Sanitary fitting
Fitting with smooth bends and no inside shoulders to block flow of waste; used to join DWV pipe.

Silicone grease
A type of synthetic grease used to lubricate faucet parts; non-petroleum base won't break down rubber parts.

Siphoning
Action occurring when atmospheric pressure forces water into a vacuum in a pipe.

Slip coupling
Used to join a new fitting into a run of copper or plastic tubing. Unthreaded, and without a center shoulder so it can slide along a tube.

Slip nut
Used on a drain (such as a sink trap). Threads onto one pipe and compresses a washer around the other to form a slip joint.

Soil stack
Large DWV pipe that connects toilet and other drains to house drain and also extends up and out house roof; upper portion serves as a vent.

Solvent cement
Compound used to join rigid plastic pipes and fittings.

Spacer
Short piece of unthreaded plastic or copper pipe cut to size; used when repairing or extending pipe. Sometimes referred to as a nipple.

Stubout
End of a supply pipe or drainpipe that extends from a wall or floor.

Stud
A vertical wood framing member; also referred to as a wall stud. Attached to sole plate below and top plate above.

Sweat soldering
A method of using heat to join copper tube and fittings.

T-fitting
Or tee. T-shaped fitting with three openings.

Transition fitting
Adapter fitting that joins pipes of plastic and metal.

Trap
Device (most often a curved section of pipe) that holds a water seal to prevent sewer gases from escaping into a home through a fixture drain.

Tube
Supply pipe that is sized nominally by copper water tube sizes.

Union
Fitting that joins two lengths of pipe permitting assembly and disassembly without taking the entire section apart.

Valve
Device that controls the flow of water.

Washer
A flat thin ring of metal or rubber used to ensure a tight fit and prevent friction in joints and assemblies. Term is sometimes used interchangeably with gasket.

Y-fitting
Or wye; DWV fitting with three outlets in shape of letter Y.

INDEX

A-B-C
Aerators, 33
Angle valves, 63
Antisiphon vent valves, 70
Augers, 29, 54
Bathtubs:
 clogged drains, 30-31
 faucets, 79
 pop-up stoppers, 46, 47
 see also Fixtures
Cartridge faucets, 41-42
Cast-iron pipes, 25
 cutting, 26
 fittings, 25, 26
 joining to plastic pipes, 16
 joints, 25, 26
 new connections, 70
Cast-iron snap cutters, 26
Cleanouts, 6
 clogs, 32
Clogged drains, 27, 28
 bathtubs, 30-31
 main drains, 31-32
 showers, 30-31
 sinks, 28-30
 toilets, 53-54
Cold chisels, 26
Cold water mains, 6
Compression faucets, 33, 34-36, 79
Conserving water, 42
Copper pipes, 17
 cutting, 17, 18, 19
 DWV systems, 17, 70
 fittings, 17-18
 joining to galvanized steel pipes, 22
 joining to plastic, 13
 joints, compression, 17, 21
 joints, dielectric union, 22
 joints, flared, 17, 21
 joints, soldered, 17, 20
 joints, union, 17, 21
 measuring, 19
 new connections, 70, 71
 shutoff valves, 75

D-E
Disc faucets, 33, 36-39
Dishwashers, 88-89
Diverters, 79
Diverter valves, 44
Drain cleaners, 28
Drains, 6
 main drains, 31-32
 see also Clogged drains; DWV systems

DWV systems, 5, 67
 antisiphon vent valves, 70
 copper pipes, 17, 70
 extending, 67-73
 new connections, 70-71
 plastic pipe fittings, 12, 13, 16
 plastic pipes, 12, 14, 70
 venting fixtures, 69
Emergency procedures:
 frozen pipes, 57, 58-59
 leaking pipes, 58
 overflows, 60

F-G
Faucet chatter, 60
Faucets:
 aerators, 33
 bathtubs, 79
 cartridge faucets, 41-42
 compression faucets, 33, 34-36, 79
 deck-mounted, 77-78
 disc faucets, 33, 36-39
 freezeproof, 80
 noise, 60
 outdoors, 80
 rotating-ball faucets, 39-41
 washerless, 33, 36-42, 79
Fittings:
 cast-iron pipes, 25, 26
 copper pipes, 17-18
 galvanized steel pipes, 22
 plastic pipes, 12, 13, 16
Fixtures:
 additional, 67, 69
 roughing-in, 73-74
 shutoff valves, 6, 10, 27, 66, 75-76
 surface maintenance, 84
Floors, 72-73
Frozen pipes, 57, 58-59
 freezeproof faucets, 80
Galvanized steel pipes, 22
 cutting, 23
 fittings, 22
 joining to copper pipes, 22
 joining to plastic, 13
 joints, 24
 joints, dielectric union, 22
 measuring, 23

new connections, 71
 threading, 24
Garbage disposers, 90-91
Gas heaters, 92, 94
Gas systems, 6
Gate valves, 63
Globe valves, 63, 75

H-I-J-K-L
Heating systems, 6
Hose bibbs, 80
Hot water heaters, 92-94
Hot water mains, 6
Hot water pipes:
 insulation, 74
House shutoff valves, 6, 10
Joist-pipe junctions, 71, 72

M-N-O-P-Q
Main drains:
 clogs, 31-32
Meters, 7
Permits, 10
Pipe cutters, 19
Pipes:
 frozen, 57, 58-59, 80
 insulation, 74
 leaks, 57-58
 locating, 68-69
 noise, 13, 57, 59-60
 pipe-sized, 11
 supports, 9
 tube-sized, 11
 see also Cast-iron pipes; Copper pipes; Galvanized steel pipes; Plastic pipes; Pipe threaders, 24
Pipe wrenches, 23
Plastic pipes, 11-12
 ABS, 12
 cutting, 14-15
 fittings, 12, 13
 fittings, screw-on, 16
 fittings for DWV systems, 12, 13, 16
 joining to cast-iron, 16
 joining to metal, 13
 measuring, 14
 PVS, 12
 shutoff valves, 75
 solvent welding, 15-16
 stacks, 70
 tapping into, 13
 water pressure, 12-13
Plumbing codes, 10, 67, 68
Plumbing systems, 5
 extending, 67-74
Plungers, 29, 53

Pop-up stoppers, 46-47
Pressure.
 see Water pressure
Professional plumbers, 67, 68

R-S-T-U
Rotating-ball faucets, 39-41
Safety precautions, 10
 soldering, 20
 toilet repairs, 49
 winterizing, 59, 80
 working on pipes, 23
Septic tanks, 56
Sewer gases, 7
Showers:
 clogged drains, 30-31
 shower heads, 44, 79
 see also Fixtures
Shutoff valves, 6, 10, 27, 66
 adding, 75-76
 house, 6, 10
Sinks, 81
 clogged drains, 28-30
 deck-mounted, 83-84
 pop-up stoppers, 46-47
 roughing-in, 73
 sprayers, 43-44
 strainers, 45-46
 tailpieces, 62
 wall-hung, 81-82
 see also Faucets; Fixtures
Soil stacks, 67
 connections, 69, 70
Soldered joints, 17, 20
Sprayers, sink, 43-44
Stud-pipe junctions, 71-72
Supply systems, 6
 new connections, 71
Sweat joints, 17, 20
Tailpieces, 62
Toilets, 48-49
 clogged drains, 53-54
 flush-valve assemblies, 52-53
 handles, 56
 inlet valves and assemblies, 50
 leaks, 55-56
 overflow, 54
 replacement, 85-87
 roughing-in, 74, 85
 running, 27, 51
 septic tanks, 56
 stopper and valve seat, 51-52

sweating tanks, 54
troubleshooting, 49
water conservation, 66
water level, 51
wax gasket, 55-56, 86
 see also Fixtures
Tools, 8-9
 augers, 29
 cast-iron snap cutters, 26
 cold chisels, 26
 pipe cutters, 19
 pipe threaders, 24
 pipe wrenches, 23
 plungers, 29
Traps, 7, 61-62
 house traps, 32
Tubes.
 see Copper pipes; Plastic pipes

V-W-X-Y-Z
Valves, 63-64
 angle valves, 63
 antisiphon vent valves, 70
 diverter valves, 44
 gate valves, 63
 globe valves, 63, 75
 leaks, 64
 pressure-reducing, 65
Vents, 7
 additional, 67
 see also DWV systems
Walls, 71-72
Washerless faucets, 33, 36-39, 79
Washing machines, 91-92
Water conservation, 42
 toilets, 66
Water hammer, 13, 57, 59
 arrestors, 59
Water heaters, 92-94
Water mains:
 cold water mains, 6
 hot water mains, 6
Water meters, 7
Water pressure, 65
 plastic pipes, 12-13
Winterizing, 59, 80